Venice Shall Rise Again

Venice Shall Rise Again
Engineered Uplift of Venice Through Seawater Injection

Giuseppe Gambolati and Pietro Teatini
Department of Civil, Environmental and Architectural Engineering (DICEA)
University of Padova, Italy

AMSTERDAM • BOSTON • HEIDELBERG • LONDON • NEW YORK • OXFORD
PARIS • SAN DIEGO • SAN FRANCISCO • SINGAPORE • SYDNEY • TOKYO
ELSEVIER

Elsevier
32 Jamestown Road, London NW1 7BY, UK 225
Wyman Street, Waltham, MA 02451, USA

Copyright © 2014 Elsevier Inc. All rights reserved

A version of this book was published in 2013 on CD-ROM by the EnviroComp Institute.

No part of this publication may be reproduced or transmitted in any form or by any means, electronic or mechanical, including photocopying, recording, or any information storage and retrieval system, without permission in writing from the publisher. Details on how to seek permission, further information about the Publisher's permissions policies and our arrangement with organizations such as the Copyright Clearance Center and the Copyright Licensing Agency, can be found at our website: www.elsevier.com/permissions.

This book and the individual contributions contained in it are protected under copyright by the Publisher (other than as may be noted herein).

Notices
Knowledge and best practice in this field are constantly changing. As new research and experience broaden our understanding, changes in research methods, professional practices, or medical treatment may become necessary.

Practitioners and researchers must always rely on their own experience and knowledge in evaluating and using any information, methods, compounds, or experiments described herein. In using such information or methods they should be mindful of their own safety and the safety of others, including parties for whom they have a professional responsibility.

To the fullest extent of the law, neither the Publisher nor the authors, contributors, or editors, assume any liability for any injury and/or damage to persons or property as a matter of products liability, negligence or otherwise, or from any use or operation of any methods, products, instructions, or ideas contained in the material herein.

British Library Cataloguing-in-Publication Data
A catalogue record for this book is available from the British Library

Library of Congress Cataloging-in-Publication Data
A catalog record for this book is available from the Library of Congress

ISBN: 978-0-12-420144-6

For information on all Elsevier publications
visit our website at store.elsevier.com

This book has been manufactured using Print On Demand technology. Each copy is produced to order and is limited to black ink. The online version of this book will show color figures where appropriate.

Contents

Acknowledgments	vii
Foreword	ix
Preface	xi
List of figures	xiii

1	**Introduction**		1
2	**A Brief History of Venice**		3
	2.1	Origins	3
	2.2	Growth	7
	2.3	The Glory	12
	2.4	Decline	21
	2.5	The Fall	27
3	**The Venetian Lagoon**		29
4	**Survival of the City**		37
	4.1	Measures Taken by the Serenissima	37
	4.2	Land Subsidence, Rise in Sea Level and *Acqua Alta*	40
	4.3	MoSE: Protection from *Acqua Alta*	48
5	**Anthropogenic Uplift of Venice by Using Seawater**		57
	5.1	The Idea's Origin	57
	5.2	The Complete Project and Pilot Projects for Venice's Uplift	61
	5.3	Safety and Stability of Venice's Uplift	69
	5.4	Can Venice Be Raised Economically, in an Environmentally Friendly Way?	74
Conclusion			79
References			81

Acknowledgments

The authors are indebted to Janet Sethre for carefully supervising the English text and actively contributing to the istorical sections. They are also grateful to Frans Barends, Allan Freeze, Devin Galloway, and George Pinder who provided helpful comments and recommendations that greatly improved the presentation. The contributions to the project of anthropogenic uplift of Venice by Domenico Baù, Nicola Castelletto, Andrea Comerlati, Massimiliano Ferronato, Mario Putti, and Luigi Tosi are acknowledged. Finally, the writers would like to thank Dr Paolo Zannetti and the EnviroComp Institute for help, encouragement and assistance in the preliminary version of this book on CD-ROM.

Foreword

Science and engineering are usually seen as a bastion of cold, rational thought, where calculation is king, and social mores play little role. The Hollywood image of a scientist is a meek soul with thick glasses who is seen in his white lab coat watching brightly colored fluids bubble through a Rube Goldberg array of test tubes, pipettes, and flasks. An engineer is generally portrayed in hard hat and work boots, at the scene of some massive construction project, slide rule at the ready. In this stereotypical world, scientists and engineers do not fall in love, do not dream of roads less followed, and do not pay homage to the glories of past civilizations.

There is only one problem with this image. It does not fit the experience of any of us who live in this technical milieu. Most of the scientists and engineers that I know have a good ear for classical music, are knowledgable about great art, and read broadly in both classic and modern literature. They tend to travel widely and revel in the history and cultural complexity of our ever-changing world. Most of the projects on which they work come with questions of societal priorities, the acceptibility of design risks, and the maintenance of social justice. Many of these projects lie at the center of the political conflicts that arise over what is the optimal mix of economic progress and environmental protection. Engineers and scientists take their responsibilities on these fronts seriously and they are often among the most thoughtfully aware of these difficult issues.

When I was first asked to join the team that was looking into the subsidence of Venice, I knew immediately that this was a project with a strong social dimension. In my first visits to Venice I fell in love with the history and culture of this exotic city. Our technical meetings took place in a villa on the Grand Canal not far from the Rialto Bridge. Huge portraits of long-dead dukes looked down on us from the walls of our meeting rooms. During coffee breaks we could amble over to the leaded windows and peer down on the vaporettos and water buses plying their trade along the canal. After long technical sessions staring at hydrogeological cross-sections of the Venetian Lagoon, or debugging the computer programs that embodied our model of the subsidence process, we could wander down to Saint Mark's Square to enjoy a glass of *prosecco* and listen to the live orchestras playing their Strauss waltzes.

When Beppe Gambolati and Pietro Teatini began to think about this book, they must have struggled with how best to put forward their technical ideas in a way that would somehow capture the complex cultural milieu that is Venice. Their proposal to help mitigate the *acqua alta* by turning subsidence theory on its head and using it to re-elevate the city is a complex and original idea. How could they present it in a way that would appeal both to a lay audience and to those that are more technically astute, while at the same time keeping the *raison d'être* of the whole enterprise, a love of this great city, front and center in the reader's mind? Is it possible to wax eloquent on computer models of antropogenic land uplift and on Tintoretto's frescos in the same breath?

What they have come up with is something one seldom sees in a single book. A technical treatise that puts forward an engineering plan, but one that is presented in the context of the art and architecture and culture of the great city that it is designed to help save.

By all means: Let Venice rise again! And let this book serve as a template for how science and engineering can serve history and culture and art. And we can hope that sooner or later, Hollywood will catch on.

R. Allan Freeze
Vancouver, Canada
June, 2013

Preface

Venice, Water, Mud

Venice is a revelation of infinitely varied beauty. Like a shipwrecked treasure chest it emerges from the mud and sheds light on the surrounding surfaces of water, silt, and stone. For Venice, mud and water are obsessively present and obsessively active, both as promise and as threat. John Ruskin, painfully astounded by the city's grace, found comfort and respite in calling it a "City in the Mud."

To call it a "showcase" or a "jewel," to construct fake Venices in Las Vegas or Shanghai, is to diminish its cruelly. Venice continues to survive the tawdry clichés and the cheap imitations imposed on it; however, survival, after all, is its constant, conscientious obsession. In the Middle Ages, a merchant community formed on the island of Torcello and raised achingly beautiful marble and fresco testimonies to faith in Jesus, faith in beauty, and faith in humanity. However, the area soon became so infested by malaria that much of the population died out. A remnant of the few surviving islanders fled to the area of present-day Rialto, from where the present city irradiated and grew.

Venice is lapped by tidal marshes, ever-shifting tongues of silt, brown and shining flats. Viscid islets gave refuge to its early inhabitants as they fled from the invading hordes. Around the city, estuaries shift in form, interweaving and overlapping, subject to the whims of tide and wind, sapped by such rivers as the Sile, the Adige, the Piave, and the Brenta.

These same natural elements that have nourished the city's beauty now threaten to destroy it. Floods attack Venice repeatedly each year. The city's surface is sinking constantly, measurably. On some days, when *acqua alta* threatens, the lurid visions of Hell that we can still contemplate in the medieval frescoes of Torcello seem to bode ill for the city's future.

Can we save Venice? In this book, Gambolati and Teatini set forth a rational, detailed, painstakingly researched plan to raise the city's level through the underground injection of sea water. The authors, experts of international status in hydrological engineering, professor and researcher at the University of Padua, offer a lucid note of hope to those of us who love Venice.

Gambolati and Teatini's voice should be heard and heeded. Whether or not the pervading political and financial powers heed their challenge and act effectively, Venice lovers should be grateful to these scientists.

Janet Sethre
Venice, June 2013

List of Figures

Fig. 2.1	*Festa della Sensa* celebrating the symbolic marriage of Venice to the Sea by Giovanni Antonio Canal universally known as "Canaletto" (1697–1768).	6
Fig. 2.2	Growth of Venice (Tosi et al., 2002).	8
Fig. 2.3	Doge's coronation on the Giants' stairs by Francesco Guardi (1678–1716).	9
Fig. 2.4	The intensive urban development of Venice historical center.	11
Fig. 2.5	The execution of Doge Marin-Faliero, painted by Francesco Hayez (1791–1882).	15
Fig. 2.6	Giorgione (1477–1510): self-portrait.	17
Fig. 2.7	Tiziano (1480/5–1576): self-portrait.	17
Fig. 2.8	Tintoretto (1519–1594): self-portrait.	18
Fig. 2.9	Veronese (1528–1588) portrayed by Palma il Giovane.	19
Fig. 2.10	Satellite image of the present Po River delta with the 1600 coastline superposed. The "Taglio di Porto Viro" shown in red diverted the ancient "Po delle Fornaci" into the present "Po di Maestra."	21
Fig. 2.11	Tiepolo (1696–1770) portrayed by Rosalba Carriera.	26
Fig. 3.1	View of the present Venetian Lagoon from the sky.	30
Fig. 3.2	Maximal marine regression during the Würm glaciations (17,000–20,000 years BP).	30
Fig. 3.3	Evolution of the Venetian Lagoon. *Source*: Modified after Carbognin et al. (1979).	31
Fig. 3.4	The Venetian Lagoon around 1000 AD as reconstructed by the Venetian historian Teodoro Viero (1740–1821).	31
Fig. 3.5	Rivers inflow giving birth to the Venetian Lagoon.	32
Fig. 3.6	*Acqua alta* in Saint Mark's Square (top) and at the market of Rialto (bottom).	33
Fig. 3.7	Example of pavement raised (Sotoportego San Cristoforo).	34
Fig. 3.8	Reading Venice submersion from paintings. (Left) B. Bellotto, S. Giovanni e Paolo (1741), detail. The two arrows give the level of the algae belt in 1741 (lower) and today (upper) as derived from the on-site observations. The painting shows that there were two front steps above the green belt. The displacement is 77 ± 10 cm. (Right) The same door today. The picture was taken during low tide and the top step of the old front stairs is just visible (green arrow). The door was walled up with bricks in the first 70 cm above the front step to avoid water penetration. *Source*: Taken from Camuffo and Sturaro (2003).	34

Fig. 3.9	A view of Palace Giustinian-Lolin painted by Bellotto in 1735 (left) and a detail of the main entrance today (right). The algae shift is 66 ± 10 cm. The main staircase is now submersed and a new wooden wharf was necessary to enter. *Source*: Taken from Camuffo and Sturaro (2003).	35
Fig. 3.10	A view of Palace Flangini painted by Bellotto in 1741 (left) and a detail of the main entrance today (right). The algae shift is 71 ± 12 cm. The main staircase is now submersed and covered by algae. *Source*: Taken from Camuffo and Sturaro (2003).	35
Fig. 4.1	November 4, 1966: the violence of the water in Saint Mark's square (top) and against the murazzi (bottom).	39
Fig. 4.2	Frequency and height of the *acqua alta* exceeding 110 cm from 1870 to 2011 (247 in all).	41
Fig. 4.3	Land subsidence of the Venetian area after 1952.	42
Fig. 4.4	Schematic litho-stratigraphic cross section (A) across the Venetian Plain and (B) below the Venetian Lagoon.	42
Fig. 4.5	Map of land vertical movement in Venice as obtained with the aid of InSAR from ENVISAT images over the period 2003–2007. *Source*: Modified after Teatini et al. (2012).	44
Fig. 4.6	Msl at Venice (Punta della Salute) from 1896 to 2010.	44
Fig. 4.7	(A) Aerial view of the Sant'Andrea fort showing the northeastern bastion (right site in the photo) collapsed into the lagoon waters. (B) Detail of the collapsed northeastern corner.	47
Fig. 4.8	Sketch of the MoSE position in the three typical phases: at rest, intermediate, and in action.	49
Fig. 4.9	Satellite view of the Lido inlet (A) in 1996 before the beginning of the construction of the MoSE infrastructure and (B) in 2009. (C) A photo of the inlet in 2011 with the planned location of the MoSE gates (red lines).	50
Fig. 4.10	(A) Constructive details of the MoSE system: (1) gate, (2) hinge, (3) housing, (4) ballast (water), (5) ballast (concrete), (6) service tunnels, (7) containment pilling, and (8) concrete piles for soil reinforcement. (B) Photo of the housing caissons during their construction at the Lido inlet.	51
Fig. 4.11	Cumulative frequency behavior of tides higher than 60 cm at Venice obtained from the sea level records over the last 40 years. The number of events above the "alarm" level is shown for the three selected relative sea level rise scenarios. *Source*: Modified after Carbognin et al. (2010).	52
Fig. 4.12	Tide forecast released daily by the Tide Center of the Venice Municipality. The bulletin can be downloaded from the Web site of the Municipality of Venice.	54
Fig. 5.1	(A) Land subsidence (feet) from 1917 to 1926 due to the development of Goose Creek oil field, Texas. (B) Land uplift (cm) at Long Beach, California, from 1958 to 1975 as a result of elastic unloading caused by water injection in the Wilmington oil field. The site locations are shown in the inset.	58

List of Figures   xv

Fig. 5.2	Case studies presented in the scientific literature where land uplift due to fluid injection into the subsurface has been observed and measured. The sites are distinguished on the basis of injection purpose: (1) Santa Clara Valley, California, (2) Las Vegas Valley, Nevada, (3) Santa Ana Basin, California, (4) Long Beach, California, (5) Angela-Angelina, Italy, (6) Taipei, Taiwan, (7) Tokyo, Japan, (8) MacKay River, Canada, (9) Pace River, Canada, (10) Cold Lake, Canada, (11) Krechba field, Algeria, (12) Lombardia-1 field, Italy, and (13) Upper Palatinate, Germany. ASR, aquifer storage recharge; EOR, enhanced oil recovery.	59
Fig. 5.3	Maps of land vertical movement (mm/year) as detected with the aid of InSAR analysis (A) for the northwest portion of Las Vegas Valley, Nevada, between April 2003 and May 2005 and (B) above the Krechba field, Algeria, in the period between 2003 and 2007. *Source*: Modified after Teatini et al. (2011a).	60
Fig. 5.4	Seismic lines and exploratory boreholes used to reconstruct the 3D geologic setting of the Venetian subsurface basin. *Source*: Modified after Teatini et al. (2011b).	62
Fig. 5.5	Seismic sequences from the profile Line01 highlighted in blue in Figure 5.4. The geological sequences are identified by different colors. *Source*: Modified after Tosi et al. (2012).	62
Fig. 5.6	(A) Plan view of the 3D FE grid. The location of the injection wells is shown. (B) Axonometric view of the 3D FE grid sectioned along the coastline. The colors are representative of the various lithotypes detected within the PLS3, PLS2, and PLC2 formations and the overlying and underlying units. The vertical exaggeration is 5.	64
Fig. 5.7	Hydrogeologic north-south cross section of the Venetian basin as obtained from the combined use of seismic data and well logs (above) and as reconstructed in the 3D FE model (below). The tops of formations PLS3, PLS2, PLS1, and PLC2 are highlighted. *Source*: Modified after Teatini et al. (2011b).	64
Fig. 5.8	(A) Well bottom overpressure and (B) injection rate versus time according to the borehole numeration provided in the inset. *Source*: Modified after Teatini et al. (2011b).	65
Fig. 5.9	Predicted uplift (cm) after (A) 1, (B) 2, (C) 5, and (D) 10 years of injection. The injection wells are marked in green. *Source*: Modified after Teatini et al. (2011b). (video5).	66
Fig. 5.10	Predicted uplift versus time at Rialto Bridge.	66
Fig. 5.11	Pilot project: layout of the injection wells and other instrumented boreholes. *Source*: Modified after Castelletto et al. (2008).	67
Fig. 5.12	Aerial view of the area of (A) San Giuliano, (B) Le Vignole, (C), Cascina Giare, and (D) Fusina. Their location is shown in the central inset. The blue triangle indicates the preliminary trace of the margin of the experimental site. *Source*: Modified after Castelletto et al. (2008).	69
Fig. 5.13	(A) Pore water overpressure (bar) averaged over the injected aquifer thickness, and (B) land uplift (cm) at the completion of the pilot project (Fusina site). *Source*: Modified after Castelletto et al. (2008).	70

Fig. 5.14	Time behavior of pore water overpressure (bar) averaged over the injected aquifer thickness, and land uplift (cm) at the center of the ideal injection triangle. *Source*: Modified after Castelletto et al. (2008).	70
Fig. 5.15	(A) Differential displacements at Venice as obtained from high-precision leveling records over the period 1961–1969. (B) Average displacement rates from March 2008 to January 2009 for a portion of Venice measured using InSAR (left) and differential displacements as derived from the displacements (right). *Source*: Modified after Gambolati et al. (2009).	72
Fig. 5.16	Expected differential displacements 10 years after pumping inception. The injection boreholes are marked in green. *Source*: Modified after Teatini et al. (2011b).	73
Fig. 5.17	Examples of permeability distribution characterized by a correlation length λ equal to 20, 100, and 1000 m. Parameter λ describes the distance beyond which the permeability values of two sites are stochastically independent. Red and blue zones are representative of permeability areas four orders sof magnitude larger and smaller than the average, respectively. The injection wells are shown by black dots. *Source*: Modified after Teatini et al. (2010).	73
Fig. 5.18	Injected formation expansion and land uplift along a vertical cross section through two wells of the pilot experiment. The results obtained with the most pessimistic permeability distribution are compared with the case of a homogeneous aquifer system. *Source*: Modified after Gambolati et al. (2009).	74
Fig. 5.19	Bathymetry of the central part of the Venice lagoon (A) as of 2002 and (B) as modified at the end of the complete injection project (Teatini et al., 2011b).	76
Fig. 5.20	Changes (in red) (A) of the lagoon bottom with elevation between − 1 and 0 m above msl and (B) of the emerged areas with elevation between 0 and 1 m above msl of the 2002 lagoon bathymetry (in blue) as predicted at the end of complete injection project (Teatini et al., 2011b).	76

1 Introduction

Venice is a miracle of beauty. Her beauty may mark your soul forever, if only you can scratch, for a moment, through the veneer of mass tourism which is both her lifeblood and her punishment.

Punishment for beauty: sublimity becomes Venice, and the price is steep.

When her first inhabitants alighted on her islands, harried by barbarian hordes, frightened, uprooted, hungry, and confused as marsh birds in a storm, they little realized what their courage and desperation and faith would help to engender.

Down through the centuries, Venice—the Phoenix, as her burnt-out-and-restored opera house is named—has arisen from the flames of history countless times. The city has always known how to bear pain for the sake of beauty, and in fidelity to beauty's messages. She has survived fires, plagues, earthquakes, wars, economic dependence on the slave trade, the Inquisition, her own corruption and prostitution, foreign tyrannies, Fascism, acts of terrorism and—down through the centuries, interwoven with all these ills—the constant ill, the constant threat whose frequency has dramatically quickened since World War II: *acqua alta*, the periodic flooding of the city during the cold season. This results from two effects conspiring together against the safety of Venice: on one hand, land settlement (or subsidence) of the city, caused both by natural compaction of the Quaternary sediments underlying the lagoon and the human-induced groundwater overdraft from the shallow fresh multiaquifer system which occurred during the 1950s and 1960s; and on the other, mean sea level (msl) rise due to global climate change.

Like modern tourism, then, Venice's waters are both her lifeblood and her punishment. Flooding threatens her each year. Its violence grows as, year after year, Venice undergoes a steady loss of elevation relative to sea level.

Today, small, stubborn handfuls of engineers are striving to stave off the city's demise. Aided by multiple, painstaking experiments based on up-to-date hydrogeological technology, and wielding statistics in light of global environmental concerns, they set forth daring but rational plans for the salvation of Venice.

This booklet presents one such salvation plan. Carefully articulated, it is based on state-of-the-art hydrogeologic knowledge and practices, and explained in terms easily within the grasp of the lay reader.

Before discussing the salvation plan in detail, we present a brief overview of Venice's history in light of her past fortunes and recent decline, and then focus on the evolution of the Venetian Lagoon. We shall consider the Serenissima's lagoon-linked projects, such as the diversion of major tributaries and the construction of protective sea walls: measures aiming to preserve that precarious lagoon environment dangling in a limbo between land and sea, which for many centuries was central to the city's prosperity, strength, and success. Our study also presents a general description of today's MoSE, the Italian acronym for Experimental Electro-mechanic Module, a mobile barrier still under construction, whose function is to separate the lagoon from the most severe northern Adriatic storm surges.

We then put forward an innovative project that could dramatically reduce the frequency of *acqua alta*, which has increased greatly after World War II because of land settlement and sea level rise. This project could mitigate the impact of the most extreme meteorological events, at only a small fraction of MoSE's cost, and with no appreciable environmental impact. It involves injecting seawater underground into a brackish sandy formation at a depth of 650–1000 m, through 12 wells encompassing Venice, entailing a uniform anthropogenic uplift of the city in conditions of absolute safety, stability, and integrity for Venice's buildings, monuments, and entire architectural patrimony. The uplift would predictably occur over a period of 10 years, and amount to around 25–30 cm. A rise of this magnitude would be capable of offsetting the sea level rise in the northern Adriatic which is expected to emerge by the end of this century due to global warming.

We strive here to anticipate and respond to the most important technical questions that may arise.

Ultimately, however, the most important question of all is implicitly addressed not to scientists and engineers, but to the public at large: will humanity today—as harried, frightened, uprooted, and confused as Venice's forefathers—find the intimate courage and the collective resources to save this city from threatened disaster?

2 A Brief History of Venice

2.1 Origins

According to legend, the earliest inhabitants of the Venetian Lagoon were a few scattered fishermen and hunters from Altino. Driven away by Attila's barbaric hordes in 452AD, they found refuge on the lagoon islands. According to tradition, the history of Venice began at noon on Friday, March 25, 421AD, when Padua's authorities established a trading post in the area. Only between the fifth and the seventh centuries, as barbaric invasions poured across the Friuli plain into Italy as through an open door, did the lagoon begin to gain any consistent population.

The early inhabitants settled in the northern part of the lagoon, on islands such as Costanziaca, Mazzorbo, and Torcello. In order to survive they were forced to adapt to highly peculiar environmental conditions. The early refugees to the lagoon islands were farmers and shepherds, but it was not long before many became fishermen.

> **An Early Gaze at Lagoon Life: Cassiodorus**
>
> From the very beginning, the forefathers of Venice were engaged in battle with the waters. At best, the proto-Venetian's life probably resembled the description given by Cassiodorus, the chancellor of Ravenna, in the sixth century AD. Around the year 537, Cassiodorus describes the islands where local gatherings of the Venetiae took refuge from the barbarian invaders. Those islands, wrote the chronicler and philosopher, "served as a safe haven for all those peoples who, with the barbarian raids that came to inflict all of Italy, had thereon established the earliest foundation for the Republic of Venice. On those islands they lived as on shoals among the waves, divided up into a number of tribes, each of which was governed by a tribune. In uniting, they came to form a marvelous kind of government." Cassiodorus, personally familiar with lagoon life, describes their fishermen's huts as "similar to the nets of aquatic birds." Boats are moored in front of their doors. While addressing a small group of such folk, he writes: "your only industrial activity is to procure salt. The areas from which it is extracted by letting the water evaporate, serve as fields."

We know little about the political and social structure of the early refugee groups during this period of relative chaos, when all the main governmental structures of Italy were languishing. We do know, however, that the centers gradually arising at Grado, Caorle, Eraclea, Jesolo, Torcello, Malamocco, and Chioggia formed and developed thanks to migratory currents shifting away from the abandoned cities of Aquileia, Oderzo, Altino, and Padua.

The lagoon population took root and expanded, excavating numerous canals to facilitate navigation, consolidating island terrain with the aid of excavated material, silting up marsh flats, and establishing hydraulic structures in wide areas of marshland in order to facilitate fishing and the harvesting of salt.

Between the sixth and seventh centuries, tiny lagoon communities eventually merged to form a province of Byzantine Italy, locally governed by "maritime tribunes," as mentioned by Cassiodorus. These magistrates, working under the Byzantine Hexarch of Ravenna, protected the local communities from Longobard invaders similar to the barbarians from whom their predecessors had fled. From that time on, this loose league of islands took shape as the foundation for the Venetian Dogado or Duchy of the future.

Narses and the Bridge of Boats

John the Deacon, who wrote in the tenth century, accepted the historical tradition which portrayed Narses as a "Father" of Venice. Of Armenian origin, the general Narses was a palace eunuch of Constantinople. Centuries before Venice existed as a city, he came to Italy in support of Belisarius and the Byzantine struggle against Ostrogoth domination. Narses fought successfully the Goths up and down the Italian peninsula. When he was assigned command of the third and final campaign against the Goths, he left his outpost on the Dalmatian coast, landed in east-central Italy and marched toward Ravenna, along the Adriatic coastal route joining the Roman provinces of Aemilia and Venetiae.

In order to cross through the Venetiae, Narses made a pact of nonaggression with the Frankish commander of the region. Whenever possible, the Franks and the Goths both avoided the coastal route now chosen by Narses, since the numerous navigable streams discharging their waters into the nearby delta made the towns impassable. Narses solved this obstacle by having his men transported in small vessels which sailed up from the mouths of the rivers and were used as floating bridges to the mainland. In this way he reached Ravenna and took the Goths by surprise.

Later on, local chroniclers would take great pride—mirroring a collective pride—in the fact of the "boats": it had been "Venetian vessels" that carried Narses to victory.

For many years following, the general headed the Byzantines' civil administration in Italy, and in 567, after the death of Justinian, he was removed from his post and called back to Constantinople. As history or legend has it, he took revenge for his dismissal and the insults accompanying it by calling down Albuin, the Longobard king, to conquer Italy.

Therein lay another reason for assigning Narses with the honor of having "fathered" the city of Venice. In 568, the year of Narses' death, Albuin waged the most durably consequential of all barbarian invasions into Italy. His invasion drove increasingly compact masses of the local populace to settle some of the islands which presently uphold the city of Venice.

For centuries, as we have seen, the Veneto area was threatened by such barbarian invaders as Odoacer in 476; Theodoric, the great Ostrogoth king (and employer of Cassiodorus), in 489; and Albuin, king of the Longobards, in 568. The peoples of the archipelago, wary of the massacres and looting which accompanied the Ostrogoths' and Longobards' descent into Italy, clambered to fortify their sites by digging canals and building a small fleet.

Three important new lagoon communities burgeoned during the times of Cassiodorus and Narses. Grado, to the north, became a seat of religious power after a Christian community led by its bishop had migrated from a devastated Aquileia, fleeing from the fury of barbaric invaders. Eraclea, between the mouths of the Piave and Livenza rivers, was the first seat of government and center of civic life. Torcello, set on an inner stretch of the lagoon, still evokes the far-off echo of its ancient splendor and commercial importance.

As they began to feel safer, the islanders began to consolidate their political structures. Each island appointed a chief, or "tribune," selected among its inhabitants. His power was restricted by the people's assembly, the *arengo*, although in theory the island governments ultimately depended on Ravenna's Hexarch. Altogether the islands formed a sort of confederation. In reality, however, each of them was a small, independent republic.

The lack of a common defense system exposed the confederation to danger and threats from all sides. In order to strengthen their alliance, in 697 the islanders decided to elect a single commander, a *dux* or *doge*. His powers were vast; potentially, at least, he was expected to hold office for life, although the people's assembly that had elected him could depose him at any time. Many early doges ended up serving only for a short time: one was assassinated, four were blinded, two excommunicated, and three deposed without penalty.

In 729 the doge Orso attempted to transfer the title to his son, and thus make the office hereditary. A revolution broke out, and Orso was killed by the enraged population. The *dogato* as originally conceived was abolished, and governance of the Republic was turned over to military captains elected yearly. However, the experiment failed amidst bloody riots between rivaling factions, so after 5 years the Venetians once again began electing a doge.

At this point, however, they decided to transfer the capital from tumultuous Eraclea to the small island of Malamocco.

By the end of the eighth century the lagoon republic was a prosperous, solid one with an excellent mercantile fleet. A flourishing maritime trade came to strengthen the original economy, based on salt extraction and fishing. Venetian galleys began to cross the Adriatic and the Aegean, the Mediterranean, and the straits of Gibraltar; eventually, from the Atlantic, they began to push as far as the North Sea.

In 810 the capital was transferred from Malamocco to Rivus Altus (today's Rialto), which ensured better defense because of the shallow sea bottom surrounding the area. Earlier on, invasion by the Franks had suggested such a defense strategy.

Based on the name of its Venetic founders, the Republic came to be called "Venetia," now Venezia.

In 900 the Republic was threatened by hordes of Huns. Although land-based, and with no experience at sea, they plundered Chioggia. Aboard their rudimental ships, they then headed for the capital. The Venetians caught them by surprise in the waters off Malamocco, sank all their boats, which were unsuited to the shallow depths, and slaughtered all the survivors. This success opened up many Adriatic ports to Venice. Numerous ports put themselves under the protection of the Republic, for defense against raids by Dalmatian and Hunnic pirates.

Under doge Pietro Orseolo II, elected in 991, the Republic extended its dominance to Dalmatia by ridding the sea of pirates. Thus the Adriatic came to resemble a Venetian lake. Pietro II won the favor of both the German and the Byzantine emperors. To gain the Church's favor as well, he built several monasteries, sending off two sons and three daughters to embrace the monastic life. One historian compared Pietro II to Pericles.

To commemorate the success of this doge—who received the title of Duke of Dalmatia, recognized even by Byzantium—the most solemn of all Venetian festivals took institutional form. This was the *Festa della Sensa*,[1] celebrating the symbolic marriage of the Republic to the Sea. From his regal galley, the Bucintoro (Figure 2.1), the Doge pronounced the ritual formula *Desponsamus te mare, in signum veri perpetuique dominii* (we marry you, oh Sea, in the sign of true and eternal dominion). During the reign of Pietro II, the structure of the state was securely consolidated and organized, supported by a powerful navy and merchant fleet. Venice thus became one of the great Western powers: numerous states and factions now sought her support and friendship.

On the death of Pietro II, the Republic celebrated an elaborate funeral and asked his son Ottone Orseolo to succeed him. Ottone's reign was sorely troubled by warring between two factions that now divided Venice and other Italian centers: one pro-Germany, the other pro-Constantinople.

Figure 2.1 *Festa della Sensa* celebrating the symbolic marriage of Venice to the Sea by Giovanni Antonio Canal universally known as "Canaletto"[13] (1697–1768).

The fierce Slavic pirates having been thrust back into their dens, Venice now had to face the threat of the Normans, who in 1081 occupied Durazzo and Corfù, possessions of the Eastern Empire. From these two ports, located at the entrance to and exit from the Canal of Otranto, an enemy might easily intercept Venetian convoys moving from the Adriatic to the Ionian sea, and block the Byzantine triremes travelling in the opposite direction, loaded down with merchandise. For Venice and Byzantium, the possibility to navigate these two seas was a matter of life or death. In 1083, doge Vitale Faliero sailed toward Durazzo and Corfù with a fleet of several hundred galleys. The Normans, led by Roberto Guiscard, were defeated by the Venetians, more numerous and better armed. The two ports were restored to the Byzantine emperor, Alexius Comnenus. In return, the doge gained such great commercial advantages that Venice's traders overshadowed all others in the eastern ports. The galleys of the Republic were exempted from paying taxes and customs rights. This was a further step toward the complete liberation of Venice from vassalage to the court of Byzantium. Venice, considered Byzantium's *fedelissima ancella*, most faithful handmaid, now became so only in words.

Venice's maritime supremacy exploited her Eastern advantages for centuries: so adroitly that, by the end of 1100, the Venetian Republic had become the dominant commercial power in Europe. Now Venice was a universally recognized hub, collecting and delivering a great variety of merchandise. From the East she imported spices, perfumes, silk, brocades, and dyes, and to the West she exported construction wood, iron, copper, silver, and salt. She also exported slaves, despite the Church's prohibition against slavery. Her ships not only supplied the ports of Italy, the Balkans, and Greece, but also those of France, Spain, and Germany. Until the middle of the thirteenth century Venice exercised an absolute, undisputed commercial monopoly.

2.2 Growth

Starting from its original settlements, Venice quickly developed as an urban structure (Figure 2.2). The intertidal areas and the shallow waters surrounding them were consolidated using *palafitte*—poles sunk vertically into the ground and preserved from rot by the sea's own alchemy—and material dug up from the excavation of canals. In this way new reclaimed land areas arose. Thus were created the 120 small islands which form today's historical center of Venice.

At first, most Venetian buildings were made of wood. After many of them succumbed to fire, however, builders started to prefer stone brought in from abandoned towns in the lagoon, such as Costanziaca, and the Dalmatian territory, which lay under the control of the Serenissima Republic.

Life and civil organization came to interact ever more profoundly with the element of water: water meant refuge, safety, nourishment, trade, income, and, at the same time, prospects for development.

Venice's economic growth outside her confines depended on her political stability inside, ensured by an aristocratic constitution innovative for her time. Venice was actually a "democratic" republic only in theory. Her laws were oligarchic in nature, particularly from the eighth century on. Although the early doges were elected by universal

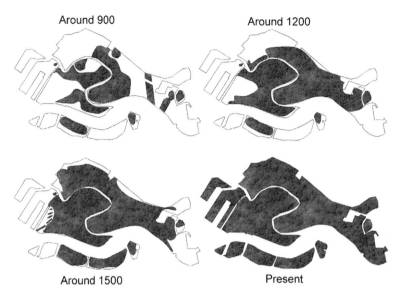

Figure 2.2 Growth of Venice.
Source: Taken from Tosi et al. (2002).[14]

male suffrage, the right to vote was later restricted to an increasingly smaller number of citizens. However, only at the end of 1100 did the doge's election become definitely institutionalized. The procedure was lengthy and complicated. In essence, the highest state organ was the Grand Council, made up of around a thousand members aged above 30, and belonging to the wealthiest and most highly aristocratic families. On election day they met, and each member drew a little ball from a box. Those who drew one of the 30 balls labeled *lector* now remained in the room, while the others left. By a similar method, the 30 electors were reduced to 9, who in turn selected 40 names among members of the Grand Council. The 40 were reduced to 12, who elected 25 names in a subsequent voting. A vote was held by which the 25 designated 9, who then were called upon to elect 43; these were then reduced to 11, who finally chose 41 voters electing the doge.

At this point of the procedure, the conclave could start. Each participant was shut up together with his servants in a room of the Ducal Palace,[2] under the surveillance of two counselors. Each voter could indicate one candidate on his ballot. The candidate receiving a minimum of 25 votes was elected as the new doge. Conclaves usually lasted 2 or 3 days, but in some cases they might last over a month, during which there was ample consumption of meat, fish, game, vegetables, cakes, spices, ice cream, wine, and liqueur. Between one voting and the next, conclave participants often played cards and chess.

Church bells tolled to announce the successful election of the doge, calling the Venetians into Saint Mark's Square, where the grand chancellor communicated the outcome of the voting. A gala banquet took place for the dignitaries of the Republic and the diplomatic service. Later on, the newly elected doge entered Saint Mark's Basilica and kissed the relics of the Evangelist. He then walked over to the Ducal

Figure 2.3 Doge's coronation on the Giants' stairs by Francesco Guardi[15] (1678–1716).

Palace to swear faithfulness to the Republic and promise obedience to the constitution (Figure 2.3). Finally he withdrew, exhausted, into his private apartments. For 3 days Venice celebrated the happy event. Crowds of commoners danced and sang. In the hall of the Ducal Palace, a great dance was held in honor of the noblemen and other notables, and there were concerts and parlor games. The doge, however, did not attend: only his relatives acted in hosting the grand event.

The earliest doges enjoyed vast powers. With the passage of time, however, these powers diminished, handed on, in part, to the Grand Council. After this institution became the highest political authority, the doge was not allowed to exercise commerce or usury, make or receive gifts, open State letters, exhibit family portraits or his family's coat of arms, or grant private interviews. Although he was entitled to be bowed to and have his hand kissed, he had to pay taxes like any other citizen, and attend mass at least thrice a week. Before taking any journey, and in some cases, before going to the theater, the doge needed to gain permission from the Grand Council. Very seldom did he emerge from the Ducal Palace. If he did so, in order not to be recognized he wore a simple tunic and hid his face behind a mask. From the thirteenth century on he was obliged to present the Venetian nobles with two wild ducks, one fat and the other thin, bred in one of his large landed estates in the Marano valley.

A doge's public life was ruled by a very rigid protocol. He was required to participate in Grand Council sessions, and to preside over the most important State assemblies. When speaking out in an official capacity, he did not use Latin, which was still the official language and lingua franca throughout Italy, but the Venetian dialect; he remained sitting, with his head covered, while the audience stood in silent attention.

His living expenses were enormous, and expenses accorded him by the State did not completely cover them. He was expected to furnish his own apartments, paying for the furniture, the carpets, and the silverware. However, the Minister of the Treasury did pay the expenses for tapestries and velvet chairs.

Like the doge, his relatives were subject to many restrictions as well. His sons and brothers were excluded from both public and ecclesiastic careers; although they were allowed membership on the Grand Council, they had no right to vote. Their only recognized privilege was priority over the other patricians. On important occasions they customarily followed the doge's procession, heading through the *calli* to Saint Mark's Square; and here the doge's corpse was carried when he died.

In earlier times, the doge's funeral was quite simple. Only after 1000 were obsequies carried out with much pomp. A few hours after the doge's death, his body was embalmed and covered with a gold mantle. The corpse was then laid on a large table between two tall, flaming candelabra in one of the palace rooms, and for 3 days it was exposed to public view. The funeral took place at sunset, accompanied by the tolling of Saint Mark's bells and the bells of other lagoon churches. An endless procession of counselors, magistrates, dignitaries, officers, admirals, and ecclesiastics, followed by an oceanic crowd, headed to the church of Saints John and Paul,[3] holding aloft the coffin, gonfalons, candles, and the relics of saints. The doge's relatives escorting the coffin wore black hooded mantles. As soon as the burial procession reached Saint Mark's the bells stopped ringing, and sailors raised up the catafalque nine times, shouting "Misericordia," have mercy on us. At the end of this rite, called *salto del morto* (dead man's leap), the procession resumed its way toward the church of Saints John and Paul, where the Patriarch celebrated the funeral mass. The ceremony was completed with interment of the corpse in the church itself. The overwhelming funeral expenses were charged to the doge's family, often forced to incur heavy debt in order to meet them.

Funeral ceremonies for the doge's wife, or *dogaressa*, were dazzling as well.

In the earlier times this woman was simply the doge's wife, with no special prerogatives. However, in the early thirteenth century, privileges and restrictions were extended to her as well. In 1229 *dogaressa* Tiepolo vowed to decline any gift except rose water, flowers, and balsam. She also promised not to incur debts and not to practice usury. During the solemnities she wore a golden mantle and covered her face with a veil. She always sat to the doge's left, upon a higher seat, and at table she was served with golden plates. She had a gondola at her disposal. As First Lady of the Republic, she often enjoyed greater popularity than the doge himself. For magnificence and wealth she equaled the *basilissa*, the Byzantine empress, and the Persian empress, while living in the most prosperous, brilliant capital of Europe and, perhaps, of the entire world.

Venice's architectural imprint was unique and regal at the same time. On festive occasions the canal waters reflected the lights and profiles of palaces ensconced in elegant silks and brocades of oriental manufacture. These were the abodes of noblemen made wealthy through maritime trade. Their names appeared in the Republic's official registry of aristocrats, and some of them, such as Morosini, Dandolo, and Mocenigo, exist even today. The most outstanding Venetian families, represented in the Grand Council, monopolized the highest state and diplomatic offices. They constituted a close-knit class of commercial origin, and had little contact with the Italian nobility of Germanic and feudal extraction.

In the thirteenth century, Venice had more than 100,000 inhabitants. At the top sat a small minority of patricians. The middle class, the backbone of the bureaucracy, was represented on the Grand Council by the Grand Chancellor who, however, had no right to speak out during assemblies. The commoners or popular class included a number of guilds, of which the arsenal workers, the Murano glass-makers, and the Burano lace-makers were the most prosperous. Often, a guild had its own patron saint and its own church or chapel.

Until 1300 no true public streets existed in Venice. Canals were the only thoroughfares. The few narrow streets, or *calli*, were often private, restricted to small neighborhoods and adjacent to the main buildings. This fact largely explains the intricate urban development of Venice, still predominant today (Figure 2.4). Just as there were no true public streets in the early times, there were no bridges either, and the only way to travel through town was by boat. After 1300, new civic ordinances were issued providing public transit leeway around the newly erecting buildings, mostly along the canals. In this way, the *rive, fondamenta*, and *rughe* were laid out for pedestrians.

Venice developed, then, partly through improvisation, with no overall, pre-established urban plan. This casualness, this lack of a planned geometry, the complete intermixing of famous monuments with popular buildings, her system of inland canals, and the city's unique nesting on the water have all joined in making Venice the incomparable jewel we know today. Due to her insular condition restricting further development, and the relative immobility of her building activities, Venice has traveled through many centuries essentially unchanged, in her unique harmony.

Figure 2.4 The intensive urban development of Venice historical center.

Venice devoted her efforts not only to trade, business, and profit, but also to pleasure and enjoyment. Many of her citizens aged between 15 and 30 were enrolled in the arbalester clubs. Each club had a number of teams formed by 12 shooters directed by a head player. Arbalest tournaments were held in December, March, and May. In summer, atop bridges devoid of any parapet which might have protected the fighters from falling into the canal below, furious boxing matches took place, egged on by a jesting public.

Another game drawing numerous spectators to the banks of the Grand Canal[4] was the so-called "Hercules performance," in which several athletes clambered to form a sort of human pyramid. This difficult exercise was performed on board a boat which at the slightest movement might capsize, and spilling its entire load into the water. The pig hunt on the Thursday before Easter, in Saint Mark's Square, was also very popular. Hundreds of pigs harried by dogs rushed headlong into a line of hunters armed with knives or axes. The square was soon turned into a horrifying public slaughterhouse, as people ran to seize parts of the slaughtered animals.

However, as we have seen, the most popular holiday was the *Festa della Sensa*, celebrating the victory of doge Pietro II Orseolo over the pirates infesting the Dalmatian coasts. On the Feast of the Ascension, the doge crossed the lagoon on board the Bucintoro; and at the entrance to the port of San Niccolò, at the Lido, he symbolically married the sea by pouring a bucket of holy water into it and then casting in a ring consecrated by the Patriarch. Afterwards the Patriarch celebrated mass in Saint Mark's, a solemn mass of thanksgiving: the *Te Deum*.

Saint Mark's Basilica[5] was erected in the first half of the ninth century. In 976 it was almost entirely destroyed by fire. Pietro II Orseolo had it rebuilt along the lines of Constantinople's Church of the Holy Apostles. Byzantine artists decorated the basilica walls with mosaics and ornaments. Ever since the ninth century, the basilica had housed the relics of Saint Mark, which two Venetian merchants had stolen away in Egypt and carried to Venice, hidden in a basket of ham and pickled pig's feet. With its striking Arab–Byzantine architecture, its Greek cross plan, and its huge central cupola, Saint Mark's Basilica resembled a missal with a cover made of precious gems. It soon became the emblem of the Republic.

2.3 The Glory

Having laid down the foundations of her power, Venice now consolidated it with an adventure known as the fourth crusade (1202–1204). Doge Enrico Dandolo led this so-called "crusade" which, instead of heading for Egypt as originally agreed by the various participating nations, turned toward Byzantium. Constantinople was assaulted and cruelly plundered. At the Doge's suggestion, the various groups of conquerors agreed to divide up among themselves the entire territory, along with its enormous wealth: a new Latin Empire of Byzantium was to be established in its place. Three eighths of the Byzantine dominion passed into Venetian hands, forming a vast colonial empire embracing all the Cyclades and most of the Aegean Archipelago, whose many islands were distributed among various Venetian aristocratic families for colonization. An almost interrupted Venetian-held chain of ports and supply points now stretched from Dalmatia to Byzantium and beyond, towards the Black Sea.

After Constantinople fell, the "crusaders" plundered an immense quantity of her gold, jewels, marbles, and works of art. It was then that the four famous horses of Saint Mark's Basilica were brought to Venice.

Her sea power having increased immensely, Venice realized that her newfound colonial and maritime grandeur were bound to fuel the poorly concealed hate and envy of the rival powers, above all the Republic of Genoa on the Tyrrhenian Sea, Venice's most feared competitor. Both Venice and Genoa had established colonies in the main ports of the Tyrrhenian. Conflict between the Genoese and the Venetians broke out frequently as each struggled to establish a monopoly in maritime traffic. Now Genoa came out victorious, now Venice. In 1298 the Venetian fleet was defeated by the Genoese near the island of Curzola. In 1379, however, when Genoese galleys sailed boldly up the Chioggia lagoon, they were blocked by the dromons of the Serenissima. Nearly all the Genoese ships sank. One year later, in Turin, a treaty was made recognizing the maritime supremacy of Venice, a condition necessary for preserving her naval and economic primacy. Genoa ended its war against Venice drained of its wealth, its domestic government greatly weakened. Gradually it fell into decline.

In the early fourteenth century, Venice was the most powerful state of Italy. However, her supremacy was bound to be ephemeral if not supported internally by a solid constitution. Although formally the doge was the commander in chief, an oligarchic aristocracy held the real power. After the middle of the twelfth century the doge shared power with a Council of Sages, or Major Council, initially made up of 35 members, and soon joined by a six-member Minor Council. The Councils deliberated on a vast range of legislative, political, and military matters. The doge, assisted by a committee of six Sages, simply ratified the decisions made by those who had elected him, and who could dismiss him at any moment.

He remained a supreme symbolic figure, however: as long as he held office, he enjoyed prestige of a sort usually accorded to the divine. His acts and gestures were ritualized; his public appearances were solemn and sumptuous, often involving a long procession of Major Council members, foreign ambassadors, and prelates of Saint Mark's Basilica. Such spectacles were designed to strike the fantasy of the commoners, perhaps in compensation for the power they had lost after the seventh century, when the lagoon islands had formed a confederation for mutual defense against the barbarian invasions.

Towards the end of the thirteenth century, the Venetian constitution was further modified, strengthening the oligarchic nature of its government. Doge Pietro Gradenigo persuaded the Major Council, which was renewed annually, to approve a law establishing, in effect, that only previous members or those whose forefathers had been members could be appointed to its ranks. Their names were inscribed in the "golden book" of the Serenissima, an official Book of the Nobility.

Another turn of the screw occurred in 1319, when yearly renewal of the Council was abolished and the number of its members increased. The Major Council now threatened to become top-heavy; many of its functions had to be delegated to more restricted committees.

A Senate was instituted towards the middle of the thirteenth century. Its powers increased progressively until the Major Council handed over to it the governance of public affairs. One of the Senate's main concerns was foreign policy. The Senate made binding decisions on peace and war, negotiated treaties and alliances, and gave instructions to ambassadors. In short, most of the Republic's authority now lay in the hands of the Senate.

A new organ formed in 1310, after a nobleman, Baiamonte Tiepolo, plotted to overthrow the State and set up a personal dictatorship in Venice. Exploiting the discontent engendered by Pietro Gradenigo's oligarchic reforms, he won over the support of several influential men. When his plot was discovered he spontaneously surrendered to the Doge. Though his life was spared, he was forced to emigrate to Dalmatia. In the wake of this incident, the Major Council elected a home security commission (*Comitato di Salute Pubblica*), otherwise known as the Council of Ten. Its members enjoyed nearly unlimited power and total independence. Initially, they were elected every 2 months, and later every 2 years. They were carefully selected after undergoing a number of tests. Their loyalty to the Republic had to be above suspicion, as did their incorruptibility. As soon as they were appointed they tended to become mysterious and inaccessible; circumspect in gesture and word, they declined to attend celebrations and public ceremonies, and met every day behind closed doors, attempting to ensure that nothing leaked out of their meetings. They could not leave Venice, and thrice weekly they met with spies and informers. In essence, they formed a secret police force. They saved the Republic by installing a regime of terror, or least, of pervasive, violent warning against any temptation to subvert the political status quo.

In 1355 the Council of Ten discovered the conspiracy by Marin Faliero, member of an eminent Venetian family. The preceding year he had succeeded Andrea Dandolo as doge. A bold, ambitious man, he was not content to remain a mere symbol. He wanted power, and like Baiamonte Tiepolo, he goaded the commoners to rebel against his own class, the nobility. He enrolled a few hundred sailors and dock hands who, on the night between April 15 and 16, were instructed to spread a rumor that the Genoese fleet was moving towards Venice with hostile intentions. According to his plan, the rumor would sow panic among the population and the nobles would gather in a body in Saint Mark's Square, where the conspirators would assassinate them. Everything was planned with the utmost care. However, shortly before the moment fixed for the coup, one of the conspirators revealed the plot to a member of the Council of Ten. He did not openly mention Faliero's name. The councilor rushed to the Doge, who fell into a web of self-contradiction, first denying the facts and then admitting his knowledge of them. When the conspirators were arrested and interrogated, Faliero could no longer conceal his involvement. The very next day he was taken before a special court made up of the Ten plus a college of 20 additional judges. He was immediately condemned to death and executed on the morning of April 17, on the Giants' Stairs (Figure 2.5): on the very spot where he had been proclaimed as Doge.

Ever since the twelfth century, Venice had fixed her gaze on the sea, whence her power had arisen and thriven. However, although she continued to preserve her maritime supremacy, she could not remain aloof from events taking place behind her back, in neighboring inland regions, as she attempted to expand her commercial policy on the

Figure 2.5 The execution of Doge Marin-Faliero, painted by Francesco Hayez[16] (1791–1882).

mainland. Her new, imperialistic ambitions drove her to hunger for ever vaster territories inland, involving her more and more fatally in conflicts tormenting Italy at the time.

The beginning of the fifteenth century marked nearly a century of struggle against the Scaligeri in Verona, the Visconti in Milan, and the Carraresi in Padua. Venice conquered Vicenza and Belluno in 1404, Padua and Verona in 1405, Udine in 1420, Brescia in 1426, Bergamo in 1428, Ravenna in 1441, and Crema in 1454. By the middle of the fifteenth century she held a territory stretching from the Po River to the south, the Adda River to the west, the Isonzo River to the east, and the Alps to the north, plus areas extending northward to parts of the Trent region, and eastward into Istria (1420) and the Dalmatian coasts.[6] Her power, security, and prosperity now seemed to have reached the height of equilibrium and grandeur.

From these inland struggles, however, the Republic emerged exhausted, and pressed on all sides by numerous enemy forces. In 1453 Byzantium fell into the hands of the Turks, seriously threatening all of Venice's colonial possessions along the Levantine coasts, as well as her economic and maritime power in the eastern seas. Moreover, as the discovery of America in 1492 opened new scenarios for maritime trade, Venice's commercial power gradually started to decline: the Serenissima lacked the energy to make the transition from a Mediterranean to an oceanic setting. Commercial powers passed into the hands of younger, more enterprising peoples, Portuguese, Dutch, and English. As her prosperity dwindled little by little, a slow, continuous decay set in. It continued for 300 hundred years or so, during which Venice consumed the huge reserves accumulated in centuries of splendor. In the fifteenth century, many considered Venice the most prosperous, magnificent, joyful city of the Italian peninsula. Her fabulous palaces facing the Grand Canal, the abodes of merchants risen to the rank of patricians thanks to their wealth, endowed the city with incomparable charm. Such palaces combined two highly refined styles in exquisite fusion, the Byzantine and the Gothic. Their interiors were as rich as the façades: the walls were decorated with mosaics, tapestries, and paintings, the vaults with frescoes, the wainscoting with golden monochromes. In erecting their mansions, the Contarini, the Gritti, the Foscari, the Tiepolo, and the Loredan had enrolled the most famous Venetian artists, and were imitated by the doges and the patriarchs in a competition for magnificence.

The school of the Bellinis helped make Venice one of the greatest artistic havens of the fifteenth century. Jacopo Bellini painted in Verona, Ferrara, and Padua, where he encountered the great master, Andrea Mantegna. After returning to Venice he opened a studio together with his two sons, Giovanni[7] and Gentile,[8] who followed in their father's footsteps, taking their place among the most brilliant painters of the time. An excellent portraitist, Gentile was invited by Mohammed II, the conqueror of Byzantium, to paint his portrait in oils. After returning to Venice, he was swamped with requests for his work. His brother Giovanni survived Gentile by 8 years to become the darling of the Venetians, the most sought-after painter of the Serenissima. Tens of his paintings of the Virgin Mary now brighten churches, palaces, and convents, as well as numerous museums. He was under contract to the Republic as official portraitist of the doges.

The art which had flourished with the Bellinis ripened and shone forth with Giorgione, Titian, Tintoretto, and Veronese, now the four chief painters of the Adriatic capital. We know very little about Giorgione (Figure 2.6). He learned to ply his brushes and mix his paints in Bellini's studio, but soon he opened his own studio and in a short time became rich. His painting reflects the influence of Carpaccio and Leonardo. He felt a strong artistic attraction toward women, whom he often painted in the nude, creating images innocent of any heavy sexual innuendo. When his last lover fell ill with the plague, he refused to leave her side: he caught the plague and died at the age of 33.

His funeral was attended by a certain Titian Vecellio (Figure 2.7), who was very few years younger. Titian, too, learned the art of painting in Bellini's studio, where he met Giorgione and became his close friend. Although he was then over 30 years old, his name was much less widely known than Giorgione's. Few artists had a slower, more laborious artistic evolution than his. It is impossible here to trace through Titian's enormous oeuvre. He died, perhaps at the age of 96, after painting hundreds of pictures. He portrayed famous personages of his time, including Alfonso d'Este I, Isabella Gonzaga, Charles V, Pope Paul III, and, in several versions, himself. Not long before his death he was commissioned to paint a Deposition for the Frari church,[9]

Figure 2.6 Giorgione[17] (1477–1510): self-portrait.

Figure 2.7 Tiziano[18] (1480/5–1576): self-portrait.

Figure 2.8 Tintoretto[19] (1519–1594): self-portrait.

as the price for his own tomb. The plague stopped him from completing that work. Hundreds of Venetians followed his solemn funeral, paid for by the Serenissima.

When Titian died, Tintoretto (Figure 2.8) was 57. He looked much older due to the misfortunes that had struck him. He was the son of a *tintore*, cloth dyer, hence the name "Tintoretto." He had been apprenticed to Titian, who after only a short time curtly dismissed him: some say out of envy, some because of Titian's aversion toward academically correct draftsmanship. After leaving the master's studio, Tintoretto studied the works of the Bellinis, Carpaccio, Giorgione, Leonardo, Michelangelo, and Raphael. He was a hard worker, eager for a challenge. After a period imposing deep sacrifice and humiliation on him, he finally won success and fame. He married a beautiful, shapely young woman who gave him eight children. He seldom went outdoors. He had no friends and said he didn't want any. He was reserved and taciturn. He used to paint for days at a time and sometimes even forgot to go to bed. Alongside Titian, he was the most important Venetian portraitist. At the age of 68, he started to paint his Paradise, a huge canvas with 500 figures largely ruined with the passage of time. This was his last effort. He died at the age of 77 and was interred in the church of Madonna dell'Orto,[10] where he had enjoyed his initial success. He was defined the "Michelangelo" of the lagoon thanks to the dramatic power of his work, and to his search for transcendent reality by way of the tangible.

In contrast, Veronese's (Figure 2.9) work focused on the domestic, familiar details of everyday life. Born in Verona, he moved to Venice when he had already learned the

Figure 2.9 Veronese[20] (1528–1588) portrayed by Palma il Giovane.

essential elements of his art. He was an amiable, friendly fellow, and salons soon threw their doors open to him. He quickly adapted to Venice's wealthy, party-loving society and, they say, was idolized by women. He specialized in decorating vaults and ceilings; his signature appears in many an elegant Venetian villa. He was fond of dogs: in many of his paintings we see a dog wagging its tail or sitting at a gentleman's feet. At the age of 38 he married the daughter of his former master, and begot 2 children. He attenuated the monotony of marriage by indulging in adulterous affairs, and was surrounded by lovers, whom he preferred to select among his clients' wives. He lived in luxury and concluded his carrier by frescoing the rooms of the Ducal Palace, destroyed by fire in 1574 and 1577. He was killed by virulent and mysterious illness at the age of 60: the bubonic plague, perhaps, although the authorities of the Serenissima did not admit as much, for to do so would have made his public funeral illegal. With Titian, Venice lost the artist who best interpreted her worldly pomp and frivolous moods.

Other "minor" arts contributed to the splendors of the Republic. Crystal masterpieces—vases, cups, goblets—were produced in Murano. A flourishing industry grew up of forging and skillfully chiseling weapons—helmets, plates of armor, shields, and daggers—that found markets throughout Europe.

The artistic renaissance accompanied a surge in the passion for letters. Classical culture was held in great esteem. Nearly every important patrician had a library, and kept his door open to poets and scholars. Ladies of high society received theologians and philosophers in their salons. The press helped spread the works of both classical and modern authors. Venice became a capital of the book industry. The books printed in Venice in the second half of the fifteenth century were often true jewels of elegance, paper quality, and printing perfection. By the end of the century, 2835 volumes had been printed in Venice. The most famous European typographer of the time, Aldo Manuzio,[11] was Venetian. After the Counter Reformation, Venice distinguished herself from other states by her religious tolerance, printing books in relative freedom.

Prophecy and Hydraulic Engineering: The Blind Sage of Adria

The city-state of Venice became intensely committed to the hydraulic engineering of its surrounding territory in the sixteenth century when, after a series of short, bloody wars, it gained dominion over the Polesine and, in particular, over the area surrounding Adria. This town was much older than the city of Venice: its inhabitants had included prehistoric peoples and Greeks and Etruscans and Romans.

The region is still known as "Polesine," meaning "land surrounded by waters." Venice was to maintain its power over the area until 1797, when the French invaded Italy.

From the 16th century on, Venice and Adria both dedicated themselves to reinforcing the banks of rivers and canals and to converting marshes into arable land, creating the so-called *retratti*. Formerly, Venice had envisioned her true mission only in connection with the sea; as the saying went, *Coltivar el mar e lassar star la terra*, cultivate the sea and leave the land to itself. The situation changed radically as the Turkish threat continued, even after Venice's victory in the Battle of Lepanto (1571), commemorated publicly by Luigi Groto, the Blind Man of Adria. Turkish armies now lurked in the Balkans, hampering Venice's importation of food supplies; Spanish and Austrian armies threatened to overrun the Serenissima at any moment; plagues came and went perennially; Franciscan and Dominican preachers exhorted the Venetians to repent of their sins (including homosexuality and excessive tolerance toward the Jews) lest they be struck down by the wrath of God.

It was in this context that the Cieco emerged: Luigi Groto, the blind poet of Adria, whom Venice had adopted as a bard. Groto had lost his eyesight at the age of 8, it was said. He was only a young boy when floods destroyed many of his family's possessions near Adria. Groto became a celebrated composer of verses, a singer, lute-player, playwright, and actor.

In seeking to attenuate the continual floods around the mouth of the Po River, the Venetian authorities consulted Groto. In confronting the menace of floods, wrote the historian Carlo Silvesti in 1735, "Many were the suggestions given by various priests and professors, but none found public approval. Finally they turned to the Oracle of the Lords of Adria: that is to their blind man, a learned man, not lacking in knowledge pertinent to such matters. He suggested forming a Taglio, cutting a canal which, beginning at Fuosa and proceeding for about three miles, might divert the waters of the Po to the […] Goro basin[…] and carry them into the sea."

The blind Oracle's solution, known as the Taglio di Porto Viro, worked where others had failed.Completed in 1609 (Figure 2.10), the canal brought immediate relief to both poverty-stricken peasants and patrician speculators.

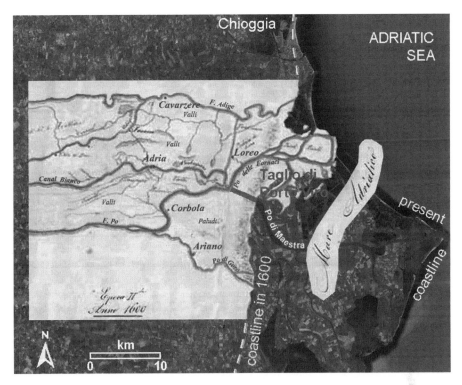

Figure 2.10 Satellite image of the present Po River delta with the 1600 coastline superposed. The "Taglio di Porto Viro" shown in red diverted the ancient "Po delle Fornaci" into the present "Po di Maestra."[21]

2.4 Decline

Her geographical position, the ability of her diplomats, and her traditional policy of neutrality and equidistance from the various power blocs had succeeded in keeping Venice sheltered from the storms hitting the Italian peninsula in the sixteenth century. Although she might agree to single political alliances as circumstances suggested, her basic independence and safety seemed ensured. Only after the consolidation of Spanish supremacy in Italy, and with the Hapsburgs' targeting of the Veneto region, did Venice decisively align her policies with France and its allies. Connecting Lombardy to Austria, the Veneto was coveted by both Madrid and Vienna, and the imperial powers made many attempts to seize control of it.

In 1615 the Austrian archduke occupied Gorizia, but he was driven back to the Isonzo River by the army of the Serenissima. In 1615 the Uskok raids against Venetian ships in the Adriatic became more frequent, but only 2 years later, with the aid of French mediation, the pirate attacks ceased and the Uskoks returned to their dens.

In the same year, the Spanish ambassador, marquis Bedmar, planned a conspiracy to overthrow the Republic from the interior. According to the plan, mercenary troops paid with his gold were to break into mutiny, blow up the Arsenal, occupy the Ducal Palace, and then proclaim Spanish sovereignty over the Serenissima. The plot was uncovered, Bedmar flew to Milan, and the Republic was saved.

During the Thirty Years War (1618–1648), which marginally involved Venice, anti-Austrian hostility stirred up again. The Serenissima entered the field alongside Piedmont and France, and shared their position in the struggle over the succession of Mantua.

Unfortunately, numerous short wars and skirmishes continued to bleed the Republic, already weakened by the huge expenses required to meet the constant challenge to her supremacy at sea. Venice was no longer "Mistress of the Seas," as she had been in the fourteenth and fifteenth centuries. The shift of maritime traffic toward the North Sea and the Atlantic, and Turkish hegemony over the Mediterranean, no longer allowed that. Her fleets were no longer intimidating; they could not compete with other powers. Her admirals were no longer invincible. Her merchants had lost the key to Eastern trade.

Venetian wealth was now concentrated in the hands of a few noblemen, who were unwilling to risk it in maritime enterprises of uncertain outcome. Capital, instead, was invested and immobilized in large estates on the mainland. This cruel reality was masked by the city's splendid ostentation. In the *calli*, in the squares, on the Grand Canal, the signs of decadence could still be easily ignored. The Venetians did not give in to pensiveness; any excuse might justify merry-making. The Serenissima was heading into its twilight as if going to a feast.

Decline was slow but fatal, and no economic or political power could stop it. The aristocracy that had once held power was torn with jealousy and personal rivalry; it had lost its class cohesion. The Senate's authority had greatly decreased. Other State services worked in a passive manner; only the police and the spies remained active.

Music and the theater impassioned the Venetians. In the early fifteenth century, with its 200,000 inhabitants, Venice was in many respects the most advanced European city. It enjoyed musical primacy. The most famous organists often played in Saint Mark's, including the Fleming, Adrian Willaert, whose theories greatly influenced his contemporaries.

In no cities were theaters as crowded and the spectacles so rich and varied as in Venice. Only those especially invited could attend the première: the Doge, the *Dogaressa*, members of the Grand Council, notables, and foreign ambassadors. In the days that followed, the theater doors were thrown open to the general public, anxiously crowding to get in. Spectators could generally sit wherever they liked, in the stalls, on the wooden stairs, in the gallery. A roar of applause broke out as the show began. The audience participated in most unseemly ways, often shouting, hissing, cursing, and throwing firecrackers. The singers were no more courteous than the spectators. They often failed to appear on schedule, stepping onstage even an hour late; they might stop the show abruptly, drink coffee during the performance, sniff tobacco, talk with the ushers, and wink at the ladies. If they thought they weren't getting paid enough they might interrupt the performance, leaving everybody in the lurch. Some stage artists earned huge amounts of money and led the life of satraps or queens, idolized by the people, sought-after and honored by popes, princes, and kings. The *castrati* were often the most capricious performers and the ones in greatest demand.

Into this bizarre and freakish world, Monteverdi launched the melodrama and fixed its rules. He worked in Milan, Cremona, and Mantua, where he composed his most celebrated opera, "Orfeo." In 1612 he moved to Venice, where he was triumphantly received and appointed master of Saint Mark's chapel of the Doges. He became internationally famous. His operas were printed and acclaimed throughout Europe. In 1631, following a period of the plague, masses of people met in Saint Mark's to attend the mass of thanksgiving composed for the occasion by Monteverdi. He died in 1643, and his solemn funeral was performed at the expense of the Republic. With him, lyrical theater lost its first maestro, and the Serenissima lost its scepter as the capital of opera.

In the second half of the seventeenth century, Venice's relations with Vienna worsened. In searching for a maritime outlet, Vienna had now set her sights on the Adriatic Sea. However, Venice's most threatening enemy was the Turkish Empire, which dominated the eastern Adriatic and Mediterranean. Conflicts between the Venetian Republic and Constantinople, the "Sublime Porte," were countless. Venice attempted to settle such controversies peacefully by paying vast amounts of ducats even when she was certain of her rights. She was afraid to initiate a war with the Sultan.

Eventually, she was forced to give up her last remaining Aegean position, the island of Crete, which she lost after a long battle in 1669. For some 15 years, Venice remained resigned to losing that last eastern rampart. However, after the Sultan besieged Vienna and was defeated there by the Polish army of Jan Sobieski in 1683, the nations forming the anti-Ottoman Christian league invited Venice to join them and offer the services of her fleet. Venice accepted on condition that Crete be restored to the Serenissima.

Command of the fleet was given to Francesco Morosini, one of the greatest admirals of all times. Morosini belonged to one of the top aristocratic "two hundred families" of Venice, and an ancestor of his had also been Doge. He earned his stripes as admiral during Crete's resistance; when defeat came to him he signed a treaty of surrender and ordered the island to be evacuated.

Once having returned to Venice, he was accused of becoming wealthy at the expense of the Republic. The accusation could not be proven, however, despite common belief that it was well-founded, and he was acquitted. He lived like a king in a magnificent palace; he loved luxury and pretty women. He was extremely ambitious, but at the same time a clever politician, wise diplomat, and, above all, an excellent general. Despite his earlier loss of Crete, under his guidance, Venice and its allies now won several dramatic victories against the Turkish empire, which was forced to sue for peace.

The peace treaty was signed in 1699 at Carlowitz, assigning Venice domination over the islands of Morea, Egina, Leucade, and Zante. Having helped to fabricate that triumph, in 1693 Morosini was elected Doge by public acclaim. In Morea, when informed of his election, he immediately asked how much money and what honors he might receive from the office once he returned to his fatherland. He remained in office for 3 years, but soon resumed command of the Venetian fleet, which had been left in the hands of timid commanders. His last journey was to the Levant, where he soon died due, in part, to kidney stones. His autopsy revealed a 6-oz kidney stone. With him, Venice lost her last famous *condottiere* and gave the last farewell to her grandeur.

The eighteenth century started out rather well for Venice. Her most recent acquisitions in the eastern Mediterranean came to join her inland territories, embracing the entire Veneto, including Padua, Verona, Vicenza, Treviso, and Belluno; the Friuli region; Brescia and Bergamo in Lombardy, and the coastland of Istria, Dalmatia, and Albania, all the way down to Corfù. Although it had shrunk somewhat compared to its most glorious times, this empire was still capable of offering Venice's 140,000 inhabitants sufficient political and economic resources, assuming that she was able to manage them.

The Turks now started a new war, and Venice failed to defend her new eastern colonies, nearly all of which she lost. Only Corfù remained, the farthest outpost in the Adriatic. Venice might have found some benefit from such downsizing if she had not lacked efficient leadership. The Venetian nobleman was traditionally a man of commerce and a man of the sea. He had little vocation for land management or large-scale industry, and he lacked a far-sighted political vision. He continued to consider his inland possessions as colonies. In the eyes of its political leaders, the Venetian Republic was only Venice, the City. The remainder of Venice's possessions was viewed as conquests, as part of a proconsulate, even though it incorporated such economically and culturally advanced cities as Padua, Treviso, Vicenza, Brescia, Bergamo, and Udine. Within the city of Venice, patricians viewed the *terrafirma* as a poor relative. They thus refused inland noblemen the right to be registered in the "golden book" of the aristocracy, which was restricted to the big-city dynasties.

After the War of Crete, however, which decimated the Venetian aristocracy, patricians from the *terrafirma* were allowed to register in the golden book by paying 100,000 ducats. This price forced some of them into bankruptcy. In 1775, 40 more *terrafirma* patricians were invited to register their names in the golden book at half the former price; however, only 10 agreed to pay. The golden book had lost its fascination, just as Venice had lost her political and economic preponderance.

The main reason for her loss was linked to the failure of economic reforms. After losing her Eastern markets, Venice should have looked for alternative commercial outlets in Austria, Germany, and Switzerland. This, however, would have required two conditions: first of all, greater freedom of exchange; and second, the abolition of corporative constraints which tended to keep salaries high, and to prevent the hiring of well-prepared foreign professionals. This second condition failed to materialize, in part, because the upper class strove to preserve the guilds, which ensured employment; by preserving them, the men of power hoped to stave off political claims and unrest, and to keep the political reins tightly in hand.

However, Venice's most clamorous failure occurred in agriculture. No sooner had maritime trade decreased then the nobles started to invest their capital in inland estates. However, they failed to approach the adventure in an entrepreneurial spirit because they had little familiarity with the land. They built splendid, rather impractical villas dramatically different from buildings on Lombard or Tuscan farms, with their warehouses and cellars.

In any case, passage of Venice's economic barycenter from sea to land entailed huge investment in estates. In 1740, 50% of mainland estate lands belonged to Venetian patricians. When sent inland, though, the old Venetian trader-admiral, who had once dazzled the world with his courage, initiative, and cleverness, turned into a listless, indolent landowner.

After the decline of Venetian prestige, espionage became a major weapon of defense for the upper class. In 1780 two patricians, Pisani and Contarini, publicly denounced the police system. Perhaps, their initiative was based on personal grounds, since they belonged to the category of *Barnabotti*, the impoverished aristocrats to whom the State often provided free lodging in the neighborhood of campo San Barnaba. Pisani and Contarini launched several radical proposals meant to reform and democratize institutions by reducing economic inequality and the power of the Ten. Public opinion supported the proposals and the popular districts witnessed a few attempts at rebellion. However, when the police suddenly caused the two *Barnabotti* to disappear—imprisoning them, presumably—nobody lifted a finger to save them. Such inertia demonstrated that the Venetian commoners, long deprived of political power, had lost interest in the institutions and were vulnerable to abuse of any sort.

Indeed, there lacked a substantial middle class in Venice which might have attempted to reform the political system. The *petite bourgeoisie* included mainly state employees strongly tied to the upper class, whose habits tended to copy and reproduce. This class formed the middle and upper bureaucracy, and defended its caste privileges just as the aristocracy defended its institutional power. Another bourgeois sector, based on commerce and the liberal professions, included a good number of Jews. It was now reduced in number and weakened by the decline of Venice as a commercial hub. The Council of Ten kept this element under constant control.

Toward the end of the eighteenth century, 23,000 beggars and 13,000 servants lived in Venice. In short, four-fifths of the population were proletarians, and showed little eagerness or illusion concerning reform.

As institutional control by the police increased in politics, it relaxed on the personal moral level. Not even the Rome of Petronius or Byzantium in its decline had experienced such a magnificent agony. The Venetian calendar was brimming with events. A prelude to Carnival started on the first Sunday of October, paused shortly for Christmas and Epiphany, resumed and then lasted until Lent. After this interruption, dictated more by physical exhaustion than devotion, the partying started up again. Any occasion was good for the revival of a Carnival-like atmosphere, including the election of a judge.

Gambling was uppermost in the authorities' concerns. Multitudes were contaminated by the passion of gambling. According to some historians, gambling lurked in the genetic-cultural makeup of the Venetians because of their traditional attitude toward risk. Now that their maritime supremacy dwindled, the risk-taking attitude found its way to the casino tables spread throughout the city. Amorous pursuits also developed in their game-like aspects. If we can believe certain chronicles of the day, words like "faithfulness," "honor," and "virtue" faded from the common vocabulary, as always happens when love is replaced by false gallantry. License was a product of the general corruption accompanying a decaying society.

During Carnival, the *tabarro*, a coat reaching down to the feet, and the *bautta*, a thick veil covering the face, equalized the sexes. The general rule was that nobody was to recognize anybody. Naturally, anonymity prompted all kinds of intrigues and affairs. The patrician could not avoid a fellow in disguise from entering his home; one can easily imagine that many a plebeian might exploit the occasion to enter an aristocratic palace. Underneath a *bautta* a nun could sally out of her convent, a lady enter a tavern. According to Pietro Giannone, one could see women of any condition—married, single, and widowed—mixing with courtesans or prostitutes, the mask making them all equal; and no indecency did they refuse to commit with the men who desired them, young or old.

The relaxed tenor of sexual morality was mirrored in painting styles, which acquired in lightness and easy grace what they had lost in grandeur. No more could one observe a touch similar to that of the Bellinis, Giorgione, Titian, Veronese, or Tintoretto. Even while maintaining technical excellence, theme-wise, painters often conformed to the modest dimensions of a provincial society that had lost its high motivation and the taste for wide horizons. Canaletto, Guardi, and Longhi were excellent masters of provincial forms. With detailed precision they concentrated on recording the minute, everyday aspects of their city.

The most representative painter of the eighteenth century was Tiepolo (Figure 2.11), comparable to Tintoretto in skill and fantasy, but not in vigor. For Tiepolo, painting

Figure 2.11 Tiepolo[22] (1696–1770) portrayed by Rosalba Carriera.

was not a deep existential a problem, not an agony. His figures, whether women of the humble classes or Madonnas, saints or warriors, brigands or cherubs, always display good health. Even Christ on the cross expresses the joy of living. Tiepolo's pronounced sensuality led him to see a world turgid and golden. He was the perfect painter for pleasure-seeking Venice which, even while foreseeing her imminent destruction, was emptying her storerooms to the bottom. A happy-go-lucky *bon viveur*, whether in the family, in his adulterous affairs, health-wise, or in pursuing his career, Tiepolo was the perfect interpreter of eighteenth-century Venice. Among his many fortunes, he had the good luck not to see the end of his voluptuous world, since he died in 1770, honored by the multitudes.

Venice became a world center of painting patronized by the government itself. Venice understood the importance of her artistic patrimony: as early as the seventeenth century she had often forbidden private citizens from selling Venetian paintings abroad. In the eighteenth century, this prohibition was translated into law, thus safeguarding Venetian masterpieces.

As mentioned in our discussion of Monteverdi, music was another art strongly encouraged by the institutions. It came to be considered a State service, analogous to charitable works. The violinists, harpsichordists, and singers who were awarded a diploma from a Venetian conservatory were hired sight unseen by great French, English, and German theaters, thanks to the prestige enjoyed by Venetian musical education the world over. Goudar wrote that among all the overtures he had listened to, only one composed by a "certain Vivaldi from Venice" was capable of expressing a symphonic quality. De Brosses observed that in Venice, music was an incredible "passion."

Eighteenth-century Venice must have been a truly delightful city. Despite their many deficiencies, we must admire its rulers who, unable to maintain Venice's prestige as an imperial capital, did succeed in maintaining her supremacy as a capital of pleasure.

2.5 The Fall

On April 7, 1797, by signing the Treaty of Leoben, Napoleon and the Austrian plenipotentiaries agreed that Lombardy would pass into the hands of the French, while Venice would be handed over to Austria, together with all her inland and Dalmatian possessions. Napoleon quickly began to liquidate Venice. A few days after signing the treaty, he mailed an insulting letter to Doge Manin ordering him to suppress the Inquisition and the Senate. The Grand Council met for the last time on May 11, its members wearing their ceremonial uniforms, their wigs, and their robes. Of Venice's past greatness as a State, this, at least, remained. The few attempts at resistance were half-hearted. The Serenissima's death certificate was duly signed and on the night between May 15 and 16, a French commissioner took control of the city while waiting to hand it over to Austria. Austria did not wait long before annexing Istria and Dalmatia all the way down to the Mouths of Kotor or Cattaro.

The final act of the tragedy took place on October 17, the same year when the final peace treaty between Austria and France was signed at Campo Formio.[12] Diplomatic maneuvers had made very few modifications in Napoleon's preliminary text. The Venetian State now ceased to exist, and became a mere province of Austria. Its territory to the west of the Adige River was annexed to France. The very next day Austrian troops entered Venice, and sorrow broke the heart of the last Doge, the elderly Manin. Mortally stricken, he fell to the ground. A stroke of the pen had cancelled out fourteen centuries of glorious history. So the Serenissima died, without honor or heroism, torn between the cowardice of a corrupt aristocracy and a fanatical illusion of democracy.

Down through the centuries, it had struggled to thrive and flower by unflinching, imaginative, scientific attention to its dependence on the waters. Our memory of such emblematic figures as Cassiodorus, Narses, and Luigi Groto reflects this incessant struggle. The struggle itself found institutional form in the Water Magistracy, which we shall discuss below.

Venice has wrestled with the waters as Jacob with the angel. A beacon of culture and of courage in facing the elements, she certainly deserved a less ignoble, humiliating defeat than the one inflicted on her by Napoleon. The military and political defeat, however, could not erase her enduring spiritual heritage. Venice will never again be a city-state, a colonial power dominating global commerce. Her potency remains, instead, in the realm of beauty. In light of that, the struggle against the waters must go on.

End Notes

1. http://www2.comune.venezia.it/turismo/feste/sensa/en_sposalizio.asp
2. http://palazzoducale.visitmuve.it/en/home/
3. http://www.aviewoncities.com/venice/santigiovanniepaolo.htm
4. http://www.aviewoncities.com/venice/canalgrande.htm
5. http://www.basilicasanmarco.it/eng/index.bsm
6. http://www.tandfonline.com/doi/abs/10.1080/03085695508592089
7. http://en.wikipedia.org/wiki/Giovanni_Bellini
8. http://en.wikipedia.org/wiki/Gentile_Bellini
9. http://en.wikipedia.org/wiki/Frari
10. http://en.wikipedia.org/wiki/Madonna_dell%27Orto
11. http://en.wikipedia.org/wiki/Aldus_Manutius
12. http://www.napoleon-series.org/research/government/diplomatic/c_campoformio1.html
13. http://www.canalettogallery.org/
14. http://www.agu.org/pubs/crossref/2002/2001GL013211.shtml
15. http://www.wga.hu/frames-e.html?/bio/g/guardi/francesc/biograph.html
16. http://en.wikipedia.org/wiki/Francesco_Hayez
17. http://www.britannica.com/EBchecked/topic/234034/Giorgione
18. http://www.britannica.com/EBchecked/topic/597229/Titian
19. http://www.britannica.com/EBchecked/topic/596682/Tintoretto
20. http://en.wikipedia.org/wiki/Paolo_Veronese
21. http://www.parchideltapo.it/taglio.del.po/E06.html
22. http://en.wikipedia.org/wiki/Giovanni_Battista_Tiepolo

3 The Venetian Lagoon

Venice was blessed by her location in the center of a unique lagoon, which functioned as a moat, making the city resemble a fortress unassailable from either land or sea. The Lagoon of Venice is 550 km^2 wide, shaped like an elongated arch (Figure 3.1). The major and minor axes are 50 km and 8 km long, respectively.

Approximately 17,000–20,000 years ago, with the last glacial phase culminating in the Last Glacial Maximum (LGM), after the Adriatic sea had dropped about 120 m below present level and much of today's sea bottom was exposed (Figure 3.2), the sea started to rise. It reached its present elevation 6000 years ago, with no important changes afterward except for small fluctuations above and below today's level. In its northward movement, the Adriatic progressively flooded the lowlands, causing considerable regression of the coastline. This period of marine ingression is known as Flandrian or Holocene transgression. Due to recent advance-and-retreat processes, several lagoons were formed, including today's Romagna–Veneto lagoons, which represent the latest phase of this slow environmental evolution.

Due to the intense, prolonged alluvial period following stabilization of the Adriatic Sea level, a great deal of fluvial detritus poured into the sea and was redistributed along the coast by waves and sea currents, forming a littoral border which outlined the primeval lagoon (Figure 3.3). It was smaller than today's lagoon (Figure 3.1), and presented a different layout of emerging land areas and inlets from the open sea.

According to a description by the Venetian historian Teodoro Viero, 1000 years ago the lagoon communicated with the open sea by way of eight inlets, compared with the three inlets present today (Figure 3.4). The lagoon's morphology closely reflected the general processes contributing to its generation over the centuries.

One such process, an immensely important one, was the activity of the lagoon's tributaries. The Adige, Bacchiglione, Brenta, Sile, and Piave rivers flowed into its basin (Figure 3.5). They caused the lagoon water to be more brackish, less salty than the Adriatic. At the same time, they threatened to silt up the lagoon and turn it into a marshland.

Such a process would inevitably have caused the lagoon to disappear, joining Venice to the mainland. Realization of this fact inspired massive intervention by the Serenissima, including diversion of all the major rivers flowing into the lagoon, and later on, the construction of thick seawalls to protect the city from being swept away by the sea. Later, we shall discuss this point in detail.

Such impressive works helped prevent the lagoon being turned into a marshland. As time passed, however, the lagoon's sea-like characteristics prevailed more and more strongly. As no fluvial detritus now entered the lagoon, the compaction of natural underground sediment lowered the level of the sea bottom. The digging of

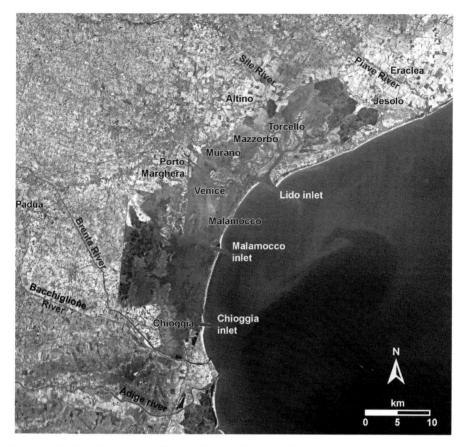

Figure 3.1 View of the present Venetian Lagoon from the sky.

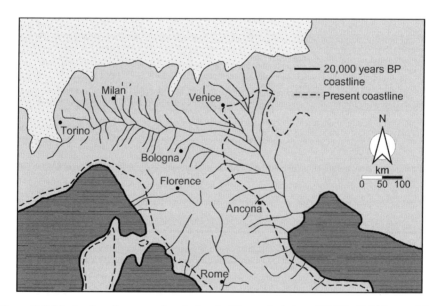

Figure 3.2 Maximal marine regression during the Würm glaciations (17,000–20,000 years BP).

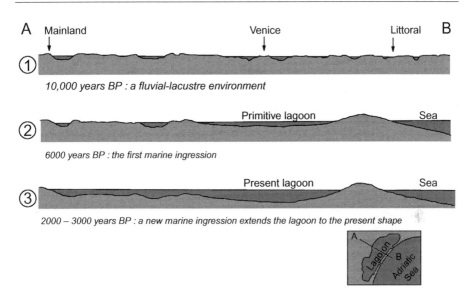

Figure 3.3 Evolution of the Venetian Lagoon.
Source: Modified after Carbognin et al. (1979).

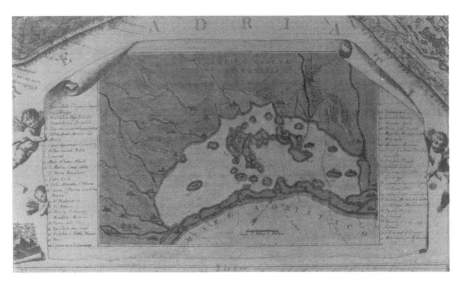

Figure 3.4 The Venetian Lagoon around 1000 AD as reconstructed by the Venetian historian Teodoro Viero (1740–1821).

canals and modification of the inlets in order to facilitate entry by ships, affected the lagoon's hydrodynamics and the erosive processes linked to it. As the marshy intertidal flats progressively shrank, productive farmlands near the edge of the lagoon were eventually submerged.

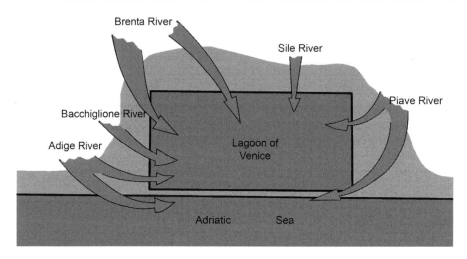

Figure 3.5 Rivers inflow giving birth to the Venetian Lagoon.

For centuries, man depleted the surrounding forests to harvest trunks and poles for the *palafitte* forming the city's foundation. During the past century, man modified lagoon areas even more severely by excavating deep new canals in order to facilitate navigation, by widening intertidal flats in order to create new industrial and urban areas, and by enlarging fish farming areas, protected by dikes. Furthermore, the intensive extraction of subsurface water considerably lowered the land level. Subsidence, in this case, was much greater than when caused merely by natural sinking.

In the end, the lagoon is smaller and deeper than it was a thousand years ago, and to a degree inconsistent with its natural evolution. Its waters are no longer brackish: salinity has increased to a degree typical of the Adriatic. In addition, even though the lagoon is an internal basin, its tides are as high as in the open Adriatic Sea; sometimes even higher, indeed, since they are amplified by the lagoon's present shape. Tidal propagation within the lagoon gives rise to the well-known *acqua alta*[1] (Figure 3.6).

The lagoon environment is further threatened by deterioration of the littoral: the lone, fragile bulwark against the sea's aggression. Though farsighted in many respects, the Venetians of old underestimated or overlooked such important physical processes as land subsidence and rising sea level. The need for *franchi altimetrici*, or height and distance rules establishing the difference between sea level and pavement level, became increasingly diffuse (Figure 3.7). In recent times, a great deal of evidence has pointed to the continual lowering of Venice's ground level: for example, the semi-submergence of entryways into buildings from the water, or the submersion of marble landing docks. Landscapes painted by Bellotto (nicknamed "Canaletto of the North") show that in the eighteenth century the algae ring was more than half a meter lower than it is today. This clearly shows that the sea has risen considerably with respect to the land level (Figure 3.8–3.10), combining the effects of both sea surge and land subsidence.

The Venetian Lagoon 33

Figure 3.6 *Acqua alta* in Saint Mark's Square (top) and at the market of Rialto (bottom).

The situation had become very clear even by the nineteenth century. In 1817–1819, Byron wrote in his ode "On Venice"[2] (1817–1819):

Oh Venice! Venice! When thy marble walls
Are level with the waters, there shall be
A cry of nations o'er thy sunken halls,
A loud lament along the sweeping sea.

Figure 3.7 Example of pavement raised (Sotoportego San Cristoforo).

Figure 3.8 Reading Venice submersion from paintings. (Left) B. Bellotto, S. Giovanni e Paolo (1741), detail. The two arrows give the level of the algae belt in 1741 (lower) and today (upper) as derived from the on-site observations. The painting shows that there were two front steps above the green belt. The displacement is 77±10 cm. (Right) The same door today. The picture was taken during low tide and the top step of the old front stairs is just visible (green arrow). The door was walled up with bricks in the first 70 cm above the front step to avoid water penetration. (For interpretation of the references to color in this figure legend, the reader is referred to the web version of this book.)[3]
Source: Taken from Camuffo and Sturaro (2003).

Figure 3.9 A view of Palace Giustinian-Lolin painted by Bellotto in 1735 (left) and a detail of the main entrance today (right). The algae shift is 66±10 cm. The main staircase is now submersed and a new wooden wharf was necessary to enter.
Source: Taken from Camuffo and Sturaro (2003).

Figure 3.10 A view of Palace Flangini painted by Bellotto in 1741 (left) and a detail of the main entrance today (right). The algae shift is 71±12 cm. The main staircase is now submersed and covered by algae.
Source: Taken from Camuffo and Sturaro (2003).

Remembered from ancient times, the phenomenon of *acqua alta* in Venice has become more frequent as the land has sunk and the sea risen. In the past, people were resigned to *acqua alta*: they struggled to adapt to the surges as to other challenges of nature. On some occasions, they succeeded in "opposing" nature, intervening in order to alter the natural evolution of the lagoon.

Modern life fostered the economic decay of Venice during the last century. The insular setting that had been a blessing in the olden times became a handicap for her development; she no longer seemed suited for modern activities. Whereas in centuries past, the city's urban structure was the key to protecting it against change, it now struggles to adapt to modern needs.

With respect to the past, Venice's source of income has radically changed, which has entailed socioeconomical decay and the impressive exodus of her people: over the last 100 years the population has been halved.

The uniqueness of Venice's environment has brought her unique problems.

During the last century, in the attempt to spur the Venetian economy, a remarkable plan for industrial and harbor development went into effect. However, it included measures detrimental to preserving the lagoon's ecological system, and the processes of deterioration became more and more alarming. The present trend toward establishing a marine environment may well threaten in the long run the very existence of the lagoon.

End Notes

1. http://en.wikipedia.org/wiki/Acqua_alta
2. http://www.poemhunter.com/poem/ode-on-venice/
3. http://www.springerlink.com/content/v66x181j88guh834/

4 Survival of the City

4.1 Measures Taken by the Serenissima

From its origin, Venice's lagoon had always protected it against invaders arriving by land. Behind its watery barrier, the city built up an empire stretching as far as the Aegean and Black seas. However, the lagoon is subject to an evolution which, by geological standards and even historical ones, is relatively rapid. Shut in between the land and the sea, it has risked being swallowed up by either one or the other. Saving the lagoon therefore has meant stabilizing the precarious equilibrium by which it dangled.

This task was entrusted to the *Magistrati alle Acque*,[1] the Water Magistrates, who enjoyed great power but shouldered great responsibility. These technicians were highly qualified not only by their competence, but also by their resolute actions. They recognized a clear priority for technical instruments over political ones. They strove to leave nothing to chance or the whim of man. In this respect, Venice was a pioneer in planned engineering.

Civic life was once governed by the *Savi*, the Wise Men, among whom the Magistrates were the most powerful decision-makers, while the Doge was a largely symbolic figure. The Councils were in charge of foreign and military affairs. Domestic politics was strongly conditioned by the delicate lagoon environment, which required a stable, authoritative regime capable of intervening quickly in its defense. In many respects, then, the hydraulician was more important than the politician.

In the sixteenth century, Venice seemed destined to follow the same pathway as Ravenna, a city similar to Venice in its origins but later transformed into a land-based town. In order to prevent such a transformation, the Venetian Magistrates saw only one solution: to divert the rivers flowing into the Adriatic. Discussions and controversy over this issue continued for decades. Many believed that with the technology then available, such a project was not feasible. The skeptics seemed to be right: in its dimension, the enterprise was no less difficult than the modern excavation of the Panama Canal. However, the Magistrates did not give in. They assumed full responsibility as the works began. No one knows how many human lives were lost to that titanic project, or how many billions of ducats. We know only that a miracle was worked thanks to the people's abnegation, the leaders' wisdom, and the bureaucrats' cognizant pursuit of their duty, while risking not only their career but also their neck (in those days the severe Venetian State was quick to recur to the gallows).

Safe from attack by land, Venice kept a wary eye to the sea. One radical, definitive measure would have been to close the three inlets, at Chioggia, Malamocco, and Lido, which connect the lagoon to the Adriatic. However, this would have condemned Venice to suffer an eternal plague. As it does today, the city lacked a sewage system; garbage was dumped into the canals; the work of dustman has been assigned to the sea, which sweeps out the lagoon by way of its three main inlets, bringing in cleaner water. Venice, then, was forced to remain clinging to the sea.

The Venetians of the Republic met such hydraulic challenges using various expedients. Undeniably, the Venetian leadership was equal to the task even during the period of decline. First, they built the *murazzi*, massive seawalls functioning to prevent the sea from creating new inlets and sweeping away the natural barrier of the Lido.

Bernardino Zendrini: Mathematics and the Murazzi

In discussing men's embattlement with the waters of Venice, we have glanced at wildly varying protagonists: from chronicler-monk Cassiodorus in the service of Goths in Italy, to the Byzantine general who defeated them; from a blind poet and Oracle to the suffering peasantry of Polesine. Not until the seventeenth to eighteenth centuries, however, does our gaze fall on a full-fledged hydraulic technician: a man of science, a mathematician; one of the most outstanding hydraulic engineers of eighteenth-century Europe.

Born in the northern mountain village of Saviore dell'Adamello, Bernardino Zendrini studied physics, astronomy, and engineering at the University of Padua. In 1735 he actively contributed to reclaiming swamplands in the Viareggio area, by designing a series of sluices with mobile gates near the shores of Lake Massaciuccoli. By separating the freshwater emitted by the lake from salt water flowing in from the sea, this row of sluices succeeded in rendering marshlands arable and in improving the conformation of Viareggio's seaport.

The necklace of sluices, different though it is technologically from the MoSE construction, resembles MoSE so closely in image that, in hindsight, it seems to herald it!

Around 1738, Zendrini, now an engineer at the service of the Serenissima, laid out plans for constructing the *murazzi*. The *murazzi* were not just mere dikes built to protect the coastline. They included stairways whose lowest steps sank deep under water in order to break the waves, dissipating their violent energy before they struck the upper part of the wall.

From 1744 to 1782, the *murazzi* were constructed along the shore between Pellestrina and Chioggia, using blocks of Istrian stone cemented with *pozzolano* mortar, made from a Tuscan clay which Zendrini himself had studied during his journeys to the Viareggio area. Istrian stone was the only stone available with a high resistance to sea salt. The clay-and-stone barriers replaced the fragile banks of *palafitte* and stones formerly erected to protect the lagoon of Venice from high seas.

The *murazzi* partly collapsed with the high tides that submerged Venice on November 4, 1966 (Figure 4.1). If Zendrini had witnessed that disaster, he might well have set out in search of a new kind of solution: one more daring and radical than static dikes.

Survival of the City

Figure 4.1 November 4, 1966: the violence of the water in Saint Mark's square (top) and against the murazzi (bottom).[15]

The Serenissima's second fundamental lagoon-linked measure was to keep the inlets open only far enough to allow for tidal flow. Each port was crossed not by a single, deep canal, but by a network of short, shallow, narrow canals which hampered entry by the sea. Like the seawaters, ships, too, found it hard to weave their way amidst the webbing of canals. Warships ended up unloading their cannons at Pola, in Istria, and merchant ships unloaded their goods at Lido, in order to reduce the pull due to their weight. However, even with their burden thus lightened, to reach the city they still required expert helmsmen able to follow the movement of waters flowing like blood in a lung, through that myriad of canal—capillaries built to dissipate the violence of the sea.

To maintain the delicate equilibrium of the lagoon, the government was willing to undergo any expense and mete out any penalty deemed useful. Anyone who sank a pole into the floor of the lagoon without permission could be imprisoned immediately for, as the Venetians said, *palo fa paluo*, "pole makes marsh"; even one pole is enough to create a small marsh. The Senate decreed a 100-ducat fine—a huge amount at that time—inflicted on "anyone dealing with the lagoon who was not a professor or knowledgeable in matters of water." It was even prohibited to talk much about the lagoon, because gossiping "created marshland": *paluo*!

4.2 Land Subsidence, Rise in Sea Level and *Acqua Alta*

Land subsidence is the lowering of the ground due to several causes, including orogenic movement, volcanic and seismic activity, sediment compaction, land reclamation, oxidation of organic soils, the dissolving of saline deposits, hydro-compaction, vibration, liquefaction; and the extraction of groundwater, thermal water, oil, gas, carbon, salt, and minerals.[2]

The Venetian Lagoon has experienced land subsidence primarily due to the extraction of groundwater, and secondarily, to the compaction of sediments deposited during the Quaternary Age, i.e., over the last million years, accompanying the deep tectonic movement of pre-Quaternary rocks. The compaction (or consolidation) of Quaternary sediments is very slow and difficult to quantify, as it can accelerate or decelerate at short time intervals. We can make an average evaluation by C^{14} dating of organic remnants found at various depths.

C^{14} is a radioactive carbon isotope produced by the organic alteration of animal or vegetable remains once found at ground level. Since C^{14} decays over time, we can estimate its approximate age based on its residual radioactivity. Using a sample depth we can calculate the corresponding rate of the land's subsidence at a given age. However, we can calculate overall subsidence values only for a period of 1000 years or more. For recent periods (say, the last 1000 years), we lack sufficient data, due to the continual reshaping of terrains near the surface. This means that though we may estimate the impact of sediment compaction on land subsidence by extrapolating average values, we inevitably risk making considerable miscalculations for short-term intervals, such as decades, or even a century.

If we may reasonably extrapolate in order to reach results relative to 20,000 years ago, we might conclude that the eastern Po River plain is naturally subsiding at the rate of approximately 1 mm/year. However, from 1952 to 1969, land subsidence as measured in Venice was in the range of 7–9 mm/year, implying that its cause was different.

The problem of flooding in Venice captured the world's attention during the exceptional inundation of November 4, 1966, when the water reached the unprecedented level of 1.94 m above msl. In the years that followed, this height has never been surpassed (Figure 4.1).

No words can adequately describe the impact of that disaster on the city.

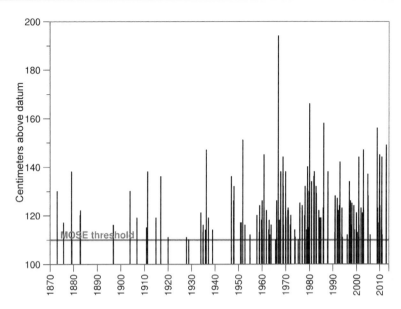

Figure 4.2 Frequency and height of the *acqua alta* exceeding 110 cm from 1870 to 2012 (254 in all).

While *acqua alta* has been known in Venice for centuries (it was mentioned as far back as 589 AD), the 1966 flood moved the entire world, spurring a worldwide movement toward "saving" Venice, recognized as part of humanity's world heritage.

According to documentary evidence, the frequency of flooding increased greatly during the second half of the twentieth century (Figure 4.2). One cause for this was land subsidence caused by the pumping of groundwater for industrial and civil use.

Before 1930, Venice seemed nearly stabilized. From 1930 to 1950 a small increase in the subsidence rate emerged. After 1950, instead, subsidence reached alarming rates; in particular, from 1952 to 1969, the land sank as much as 11 cm in Venice and 14 cm in the nearby industrial area of Porto Marghera (Figure 4.3). These are low values compared to those reported for Mexico City and Long Beach, CA (8–9 m),[3,4] or even for other Italian cities, such as Ravenna[5] and Bologna[6] (more than 1 m). However, in view of Venice's setting, even a few millimeters of subsidence may crucially affect the city's survival.

An analysis of other possible causes for the increase in land subsidence showed that land reclamation, the load exerted by construction, and the extraction of gas from the nearby Po River delta had only negligible influence on the subsidence of Venice's terrain. It also showed that the influence of natural compaction on the ongoing settlement process was quite limited. A major cause, instead, was found to be the extraction of groundwater from a 300-m-thick aquifer system underlying the lagoon (Figure 4.4A).

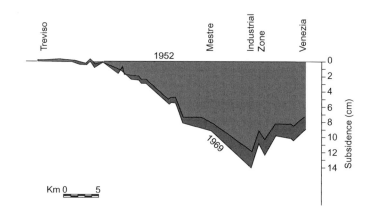

Figure 4.3 Land subsidence of the Venetian area after 1952.[16]

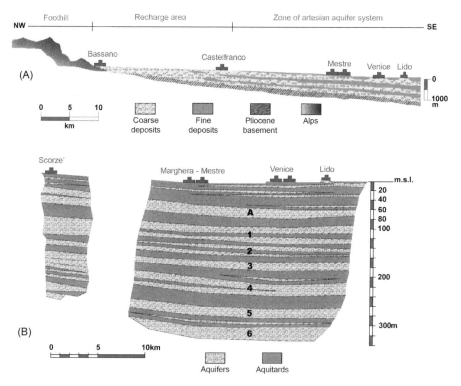

Figure 4.4 Schematic litho-stratigraphic cross section (A) across the Venetian Plain and (B) below the Venetian Lagoon.

In modern times, before 1930, the city's demand for water was satisfied by an aqueduct tapping into an artesian aquifer quite far from Venice. In past centuries, Venice obtained drinking water from covered rain cisterns located in its squares or *campi*.

Until 1930, the extraction of local water was presumably quite limited. After 1930, as the first industrial plants arose at Porto Marghera, on the mainland 7 km to the west, water consumption increased, peaking after 1950 with the intense postwar industrial development. In 1969 the pumping rate at Porto Marghera was $0.5\,m^3/s$, while in Venice it was merely 0.01. The extraction of groundwater caused a fall in pore pressure within the pumped aquifer equal to 2 bar at Porto Marghera and 1 bar in Venice.

As a major consequence, the sandy aquifers and the interlying clayey aquitards (Figure 4.4B) compacted causing a lowering of the ground surface; and the land sank. On the average, Venice subsided 11 cm due to human intervention.

In 1973 a mathematical model was constructed showing that the most inexpensive, effective way to stop land from sinking was to stop pumping water from wells.[7] This remedy would quickly restore the field of flow in aquifers, and arrest subsidence. Spurred by such studies, the Municipality of Venice prompted completion of a new aqueduct—already in the planning stage—and construction of a pipeline supplying the industrial area with water pumped from the Sile and Brentella rivers flowing near the Venetian Lagoon. At the same time, the *Genio Civile* or civil engineering bureau of Venice, like Venetian authorities of old, issued a decree prohibiting new wells and ordering the closing of the old wells as soon as the new aqueduct became operative.

These measures proved highly effective. By 1978 the well water level beneath Marghera and Venice had risen nearly to the height reached prior to depletion of the aquifers. The large cone of diminished pore pressure disappeared quickly, as groundwater flowed out again freely from numerous boreholes. The diminishing anthropogenic land subsidence now inversely tracked the increasing pore pressures. After the subsidence rate slowed down, the land soon stopped settling altogether; an uplift of 1–2 cm was observed in the late 1970s, consistent with predictions according to the 1973 mathematical model. This modest rebound, accounted for by the unrecoverable compaction of the clayey aquitards (Figure 4.4B), represented only 15–20% of the subsidence which had formerly been caused in Venice by the pumping of groundwater. Presently, the ground surface in the city is almost stable, i.e., it is subsiding at the natural rate of 1–2 mm/year, as clearly revealed by Interferometric Synthetic Aperture Radar (InSAR[8]) analyses (Figure 4.5).

Acqua alta[9] is conventionally considered to occur whenever the water elevation in the lagoon exceeds 110 cm above the msl as recorded in 1897. However, some parts of the city are flooded even at average levels lower than 110 cm. For instance, Saint Mark's Square may be flooded at a recorded level of 80 cm, while at 110 cm 14% of the city lies under water.

Chronicles reveal that *acqua alta* has been known in Venice for more than ten centuries. Only in recent times have precise measurements of the water level been recorded and kept on file (Figure 4.2). In the distant past, the expression *acqua alta* ostensibly referred only to exceptional flooding; sometimes chroniclers exaggerated dramatically: according to several accounts, on November 10, 1332, the water rose to 5.75 m above street level!

Figure 4.2 shows that the frequency of *acqua alta* has increased greatly since World War II.[10] This is due both to land settlement and to surges in sea levels.

Figure 4.6 shows the sea level's rise over the past century in Venice as it relates to ground level, illustrating the cumulative effect of the two phenomena. Roughly speaking, we see here that of the overall lowering of land levels, amounting to 23 cm

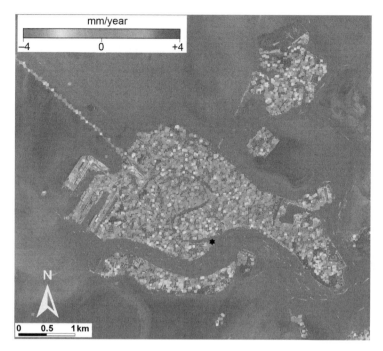

Figure 4.5 Map of land vertical movement in Venice as obtained with the aid of InSAR from ENVISAT images over the period 2003–2007.[17] (For interpretation of the references to color in this figure legend, the reader is referred to the web version of this book.)
Source: Modified after Teatini et al. (2012).

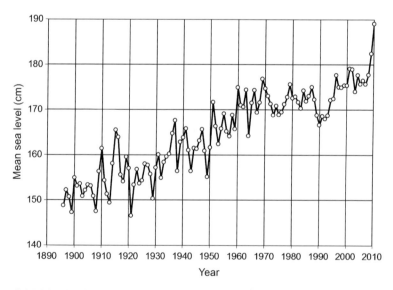

Figure 4.6 Msl at Venice (Punta della Salute) from 1896 to 2010.

Monitoring the Earth's Movement by Radar Echo: InSAR

In the last few decades, scientists have developed the ability to measure millimeter-scale displacements of the earth surface by comparing satellite images referring to intervals ranging from several years to several days. Powerful radar image processing, generally referred to as Interferometric Synthetic Aperture Radar (InSAR), has thoroughly modified our way of measuring the changes of the earth's shape over time.

Much like sound waves, radar waves carry information that echoes from distant objects; the time delay of the echo corresponds to the object's distance. "Radar," an acronym for Radio Detection And Ranging, transmits an electromagnetic microwave signal which, reflected off the ground surface, is received by satellite antenna, producing a digital image of the ground scanned. Microwaves can pass through clouds and precipitation, and function day or night, no matter what the weather is.

Each pixel, or back-scattering reflector, in an InSAR image, encodes a complex number whose amplitude corresponds to the intensity of the returning radar energy, and whose phase is a fraction of the wavelength. In the InSAR method, developed in the geophysical sciences in the late 1980s, two or more images taken from very close orbital positions at different times are combined by "interferometric processing" in such a way as to exploit the phase difference among the different acquisitions. This difference is related only to earth surface displacements which have occurred between one image acquisition and another, once the surface topography's effect is removed. In this way, displacements can be accurately evaluated.

Interferogram phase noise ("decorrelation") restricts the use of InSAR. Important causes of phase noise are temporal and spatial decorrelation. The former is due to the change of reflectors in time, as in the case of densely vegetated areas. The latter is due to the slightly different viewing positions of the orbits. Moreover, the delay in phase propagation of the radar signal through the atmosphere may be cause for error, particularly in hot, humid climates. Reflector density, and hence the measurable pixels/targets, tends to be extremely high (even exceeding 1000/km^2) in urban and deserted areas; it is reduced in farmlands, where buildings, poles, and other structures are more sparsely scattered over the landscape; and it drops to zero in densely vegetated natural environments, even preventing researchers from applying this extraordinarily advanced technology.

As InSAR measures relative displacements, like topographical leveling, in order to convert the displacement phase into an absolute displacement, we need to define a reference point. Quite often, the scene viewed from above allows us to identify stable points. Sometimes, leveling surveys or permanent GPS (global positioning system) stations may provide the reference point.

InSAR has recently produced impressive results in monitoring land displacements due to aquifer pumping and recharge, development of gas/oil reservoirs, geothermal groundwater production, natural consolidation of Holocene and Quaternary deposits, geological CO_2 sequestration, underground CH_4 storage, fault activation and fissure generation, earthquake-associated movements, motion of ice sheet, and volcano uplift.

over 1900–2000, 11 cm was directly due to a surge in sea level and 12 cm to land subsidence.

We shall now stop for a moment to consider the origin and magnitude of *acqua alta*. It occurs almost exclusively in fall and winter, when the area is most subject to storm surges, a basic factor in generating the phenomenon. The most critical episodes of *acqua alta* emerge when the astronomical tide and the meteorological tide combine, producing a cumulative effect: the astronomical component alone never gives rise to the event.

The sea level of the Adriatic depends on the astronomical and meteorological tides. The former is caused by attraction between the sun and the moon; oscillations depend on the moon's position in relation to the sun. The highest rise in water level with reference to the 1897 msl in Venice (measured at Punta della Salute) is around 70 cm. Since today's msl is 26 cm higher than the msl for 1897, the actual maximum rise in water level above the current msl is around 54 cm.

However, all the measurement levels mentioned in this study refer to the 1897 msl. For there to be *acqua alta*, the additional factor of meteorological tide needs to occur. Meteorological tide results when wind and barometric pressure affect the Adriatic during a storm surge. During a cyclone, which typically moves eastward, the sirocco wind drives the water toward the northwest along the axis of the Adriatic, accumulating seawater in the northern part of the sea. Simultaneously, the difference between barometric pressure in southern and northern Adriatic causes additional variability in surface water levels, with an increase in the northern area ranging from 10 to 15 cm, depending on the actual strength of the cyclone.

Finally, a third player may join the game: the *seiche*, a surface oscillation of the Adriatic which acts as a sort of resonant cavity, and is normally activated by a passing cyclone.

In short, the unfortunate combination of these three or four effects—astronomical tide, barometric pressure, wind, and *seiche*—may raise the water level in the city, causing an episode of *acqua alta* which usually lasts from 3 to 4 h.

The rise in water level in Venice may differ from that observed in the northern Adriatic or at the inlets. The difference depends on the shape and depth of the lagoon. As we noted earlier, the lagoon has undergone considerable modification down through the centuries: for example, with the diversion of its major tributaries.

The lagoon underwent important changes during the past century as well.[11] An industrial area arose, with its related port. This entailed the modification of internal navigation routes. Economic and industrial development burgeoned thanks to the vast spaces available near the lagoon; as of old, new land areas were created by using fill material obtained from the excavation of canals; and new building foundations were laid on the sandbanks now left dry at low tide.

Lagoon areas used for fish farming, the *valli*, further limited the surface area and volume capacity available to seawater inundation, and in so doing, produced effects similar to those of the new industrial areas.

Tidal patterns are influenced by lagoon morphology. The incoming waves are split by the various canals; the shallows exert a braking action; the banks deflect the

Figure 4.7 (A) Aerial view of the Sant'Andrea fort showing the northeastern bastion (right site in the photo) collapsed into the lagoon waters. (B) Detail of the collapsed northeastern corner.

water's impact. All such actions either attenuate or amplify the height of the original waves, change the velocity of currents, and delay tidal propagation.

Awareness of the consequences caused by modifying the lagoon's shape is crucial to a city used to living on the water and at water level. Consider the well-known example of Sant'Andrea, the fort located at a bend in the Lido canal where it enters the lagoon. After the Lido canal was deepened in order to accommodate more intense shipping activity, the current began to run faster, eroding the fort's northeastern corner, which collapsed in June 1950 (Figure 4.7).

Sirocco Winds and Adriatic Sea Basin are Involved in the Flooding of Venice

Tidal predictions of sea level may differ from levels actually observed, due to the effects of weather. The atmosphere fundamentally affects the ocean through wind stress exerted on the water's surface. The tractive force of the wind drags the water in the direction of the wind itself. As water accumulates and impacts on the coastline, the sea level increases rapidly. The overall increase in water level may give rise to severe flooding, especially if a positive storm surge coincides with the peak of the astronomic tide. Meteorologically induced storm surges are global phenomena occurring on differing scales and frequencies over all coastal areas. The most severe surges are caused by tropical storms, which affect areas such as the southern coast of the United States (Gulf of Mexico) and its east coast, the Indian and Pakistani coast of the Bay of Bengal, and Japan.

The astronomical tide in the Adriatic Sea normally ranges from 25 to 80 cm, and surges of similar or larger magnitude occur regularly. Meteorology in the Adriatic region is highly variable in terms of space, and can change quite rapidly in terms of time, due to the high mountain ranges which border the sea on three sides. The Adriatic is an elongated basin, approximately 800 km long and 200 km wide, which communicates with the Mediterranean Sea through the Otranto Strait. Meteorological disturbances in this region from October to March are frequently severe, and coastal areas around the northern Adriatic Sea, including the city of Venice, have often been threatened by dangerous coastal floods. Meteorological conditions that contribute to a higher-than-expected sea level along the coastline of the northern Adriatic usually involve the passage of low pressure systems or cyclones. A deep low-pressure cyclone located west of the Adriatic basin induces a strong pressure gradient, with a southeastward push along the major axis of the basin. Channeling effects caused by the long coastal mountain ridges produce a strong warm southerly wind known as scirocco or "sirocco." The term might be related to the Arabic شرقي (sharqī), i.e., "eastern," indicating an easterly wind. Sirocco winds blow along the whole length of the Adriatic, resulting in water being dragged and amassed in the northwestern end.

Sirocco speed ranges between 35 and 55 km/h, reaching a maximum along the Croatian coast, in the central part of the basin. The average speed in front of the Venice Lagoon amounts to 16 km/h. The average duration of continuous gale-force winds during a sirocco event is 10–12 h, with rare occurrences reaching 36 h. During the most famous sirocco event, that of November 4, 1966, gusts at Venice reached a dramatic maximum of nearly 100 km/h.

4.3 MoSE: Protection from *Acqua Alta*

The need to defend Venice from *acqua alta* emerged dramatically with the flood of November 4, 1966, when the city was completely submerged for a full day, with waves more than 1 m high impacting the ancient buildings (Figure 4.1).

Special Law no. 171/73, dated 1973, declared the defense of Venice from high tides to be a "national interest priority." The long-debated choice of the best solution for the city's survival began in those years.[12] In 1975 the Ministry of Public Works issued an international call for tender in order to define the most appropriate defense system. In 1982 the Council of Public Works formulated a series of recommendations, including the requirement that a major experimental phase be implemented. By law no. 798/84, the Italian State established a single consortium, the *Consorzio Venezia Nuova* (CVN), in charge of all activities performed in the defense of Venice, including preliminary studies, experimentation, design, and construction. This law also required that morphological damage be avoided, and that port activities, fishing, and the landscape be preserved.

During the following 5 years, CVN analyzed various defense hypotheses,[13] ranging from the protection of built-up areas through local forms of intervention (e.g., the raising of pavements), to modification of the lagoon's physical structure (e.g., the opening of reclaimed areas), and the realization of permanent or temporary measures at the lagoon inlets. The final solution, which was approved in 1992 by the Committee for Policy, Coordination and Control (the so-called *Comitatone*, i.e., big committee) with Special Law no. 139/92, consisted of a system of mobile gates, or barriers, enabling temporary closure of the three inlet channels. Project MoSE[14] is based on a row of 78 concealable flap gates installed on the inlet beds.

Associations for environmental protection strongly influenced the development of this project.

Planners also considered a similar solution, applied to the Rotterdam channel. They investigated the project in depth, but dismissed it on the grounds that—among other concerns—it would involve excessively high management and maintenance costs.

The approval of MoSE has stirred dramatic controversy. Recent studies warn that with especially severe weather conditions, the mobile gates would not assume a stable configuration, and would fail to prevent seawater from entering the lagoon. In 1998, on the other hand, a panel of international experts engaged to evaluate environmental impact reported to the Minister of Public Works that MoSE would be effective, and would not cause large-scale impact. In December 2001 the *Comitatone* gave their go-ahead for the MoSE project design, which was approved in November 2002 by the Venice Water Authority as well. During that month, the Interministerial Committee for Economic Programming (CIPE) financed the first payment of MoSE, amounting to 450 million euros. After rejection of the appeals made against MoSE by WWF (World Wildlife Fund) and Venetian government authorities, in 2004 CIPE funded a second installment, of over 700 million euros. Nine installments had been funded by the end of 2011, amounting to a total of nearly 4 billion euros, corresponding to approximately 75% of the total cost, which is expected to exceed 5 billion euros. Work is proceeding at the three inlets in simultaneous coordination; approximately 65% had been completed as of March 2012. According to a very optimistic prediction, MoSE is scheduled to become operative in 2014.

Each lagoon inlet will be provided with a series of adjacent flap gates. Normally, the floodgates are filled with water, and rest horizontally hinged on the inlet bottom. They are rotated up to an angle of 45° to prevent seawater from entering the lagoon any time the elevation of the Adriatic Sea is forecast at 110 cm or more above the official Italian datum (Figure 4.8). Venetian scientific institutions have agreed to fix a level of 110 cm as optimum, considering the present sea level, but this could be modified in the future, should differing needs and conditions emerge.

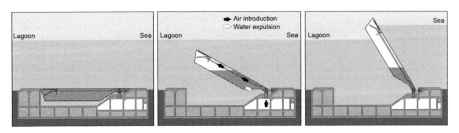

Figure 4.8 Sketch of the MoSE position in the three typical phases: at rest, intermediate, and in action.[18]

Each gate consists of a box-like steel structure that, when filled with seawater, rests inside a special recess created in the foundation structure. By introducing compressed air, water is expelled from the gate until it lifts up and assumes a tilted position. Air is then injected continuously as long as the seawater level continues to rise, so as to maintain the difference between sea level and lagoon level, up to a maximum of 2 m. The number of gates will vary depending on the width of the lagoon inlet: 18 and 19 floodgates are foreseen for the Chioggia and Malamocco inlets, respectively. As the Lido inlet is about twice as wide, an artificial island has been created in the center of the inlet, from which a series of 20 and 21 gates will connect, respectively, with either side of the inlet (Figure 4.9). In addition, the gates' dimension will differ given the different depth of the inlets and the difference in wave heights. Gate height will range between 18.5 m (at Lido) and 29.6 m (at Malamocco). At Lido and Chioggia, respectively, gate thickness will range between 3.6 and 5 m, and weigh between 200 and 300 tons. However, gate width is consistently 20 m, in order to facilitate maintenance and transport operations, and to minimize any loss of watertightness which might result from failure of a single gate. The distance between two adjacent gates will be approximately 20 cm, a width sufficient to prevent gates

Figure 4.9 Satellite view of the Lido inlet (A) in 1996 before the beginning of the construction of the MoSE infrastructure and (B) in 2009. (C) A photo of the inlet in 2011 with the planned location of the MoSE gates (red lines). (For interpretation of the references to color in this figure legend, the reader is referred to the web version of this book.)

from touching each other when they are raised, even in the presence of normally acceptable differential settlement, or when they are removed for maintenance. At the same time, this distance will be sufficiently small to limit the quantity of seawater entering the lagoon when the gates are in use.

The gates will be linked to the main foundation structure through a hinge-connector caisson which serves to secure the gates at foundation level while allowing for their rotation, the passage of air for maneuvers, and the anchoring and disengagement of the gate from the foundation structure for maintenance without any need for divers (Figure 4.10). The foundation will house the gate during normal tide conditions. The concrete caissons are enormous: 10.5 m high, 20 m wide (the same width as the gates), and 30–50 m long, depending on the inlet depth (Figure 4.10).

Three independently accessible galleries, located inside the caissons, will be used by personnel in charge of managing and maintaining the system. The housing caissons will be placed inside a 10-m-deep trench excavated in the inlet bed. In order to restrict to 3 cm per 60 m the long-term differential displacement of any individual element or of two adjacent elements, the subsoil will be reinforced by 20-m-long concrete piles arranged on a 3 × 3 m grid (Figure 4.10).

Together with the artificial island at the Lido inlet, the MoSE system has significantly altered the shape of the inlets themselves (Figure 4.9A and B). The jetties have been reshaped to accommodate the abutment connecting the gate rows to the littoral strips, as well as the refuge haven and the lock allowing vessels to enter or leave the lagoon when the gates are closed. A section of the inlet channel bed at the point where the gates will be installed has been protected by stone material laid on geotexile to prevent erosion on both the sea and lagoon sides. Off-shore curved breakwaters 500–1300 m long, with a maximum height of 3–4 m above msl, have been constructed outside the jetties to increase hydraulic resistance and reduce tidal flows.

Although MoSE's implementation is now significantly advanced, a heated debate is still going on concerning the environmental effects of closing the lagoon inlets. According to CVN results, when the mobile barriers are not operational there will be

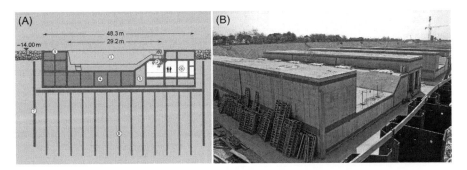

Figure 4.10 (A) Constructive details of the MoSE system: (1) gate, (2) hinge, (3) housing, (4) ballast (water), (5) ballast (concrete), (6) service tunnels, (7) containment pilling, and (8) concrete piles for soil reinforcement. (B) Photo of the housing caissons during their construction at the Lido inlet.

no reduction in the volumes of water exchanged between sea and lagoon. The lagoon inlets will be disconnected from the sea only when tides over 110 cm high are forecast, a situation that presently occurs an average of 4–5 times per year; though with significant variation, ranging from a minimum of zero events (Figure 4.2) to a maximum of 18, as recorded in 2010. Closure duration for each single event is expected to range between a few hours to a couple of days. Hence, closures under the most long-lasting unfavorable conditions could produce effects comparable to those experienced in the lagoon during neap tides, when there is slack water. Because the lagoon has a high capacity for self-purification, the water quality deterioration induced during gate closure would be cancelled out over a few tidal cycles.

Similar studies performed by other researchers have provided results that contrast somewhat with those presented by CVN, implying consequences both local (around the inlets) and lagoon-wide. A significant increase in current velocities would directly result from the new structures at the inlets, and from their new depths, causing possible erosion of the old depositional fans outside the inlets. The off-shore breakwaters in front of the inlets could deviate the seaward tidal currents, with an identifiable risk of directing the contaminants transported by lagoon waters toward the coast. Moreover, circulation within the lagoon might be affected as well, causing an enlargement of the Lido subbasin—i.e., the portion of the lagoon exchanging waters by way of the Lido inlet—at the expense of the Chioggia subbasin. A final important observation focuses on the probable, correlated rise in sea level. Projections made to date present a vast range of possible conjectures, suggesting that floods higher than 110 cm above datum could increase from the present 4–5 times per year to 21–250 times per year (Figure 4.11).

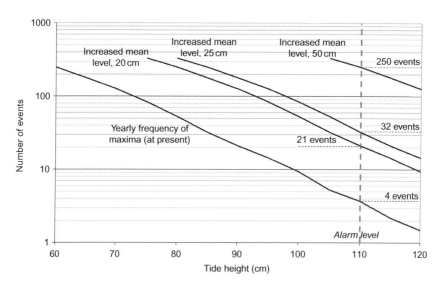

Figure 4.11 Cumulative frequency behavior of tides higher than 60 cm at Venice obtained from the sea level records over the last 40 years. The number of events above the "alarm" level is shown for the three selected relative sea level rise scenarios.
Source: Taken from Carbognin et al. (2010).[19]

Hydrodynamic Modeling of Water Levels in the Venice Lagoon

The Venice Lagoon, situated in the northwest corner of the Adriatic Sea, encompasses an extraordinarily complex network of channels and shallows. A few major channels, approximately 15 m deep, cross through an area of very shallow water averaging 1 m in depth. The presence of many islands and semi-submerged tidal marshes adds to the complexity of the circulation pattern. Water level and water velocity within the lagoon, its so-called "hydrodynamic" or "circulation" variables, depend largely on sea conditions at the three inlets connecting the lagoon to the Adriatic. Circulation generated by the Adriatic tides and local winds is also influenced by interaction with the lagoon bottom. Due to the complicated lagoon morphology and sea-floor topography, we can fully understand and reliably predict lagoon hydrodynamics only with the aid of mathematical models.

Since the 1970s, considerable effort has been made to model the hydrodynamics of the Venice Lagoon. In the early 1970s, researchers produced a preliminary, one-dimensional model focusing on the dynamics in the channel network. This model was quite successful in describing general flows and tidal propagation. However, its one-dimensionality limited its application to problems centering on the water-level response of the lagoon.

Since the mid-1970s, several two-dimensional numerical models have been developed to predict tidal levels and water velocity in the inner water body. Because of the lagoon's shallow depth and the relatively high tides at its inlets, virtually no water stratification can occur, so that we may consider the hydrodynamic system as uniformly mixed. These and similar models thus solve so-called *shallow-water* equations: i.e., two-dimensional, depth-averaged simplifications of the actual three-dimensional water flow, where vertical velocity is not taken into consideration. Increasingly refined regular or unstructured grids have been used in recent years to represent lagoon morpho-bathymetry, based on various assumptions meant to account for the dissipation of wave energy at the steep boundaries between channels, tidal flats, and salt marshes.

Hydrodynamic modeling is presently used for both short-time daily predictions of tidal levels in the lagoon (hence the possibility of an *acqua alta* event and its expected height), and research focusing on the long-term evolution of the lagoon environment. Three times daily, the Tide Center of the Venice Municipality provides a two-day tide forecast, and communicates predicted water levels to citizens using the Web (Figure 4.12 shows an example of the document released by the Municipality). Lagoon hydrodynamics is simulated by the SHYFEM model, developed by the Italian National Research Council. SHYFEM solves the shallow-water flow equations for the entire Mediterranean Sea, including the Adriatic and the Venice Lagoon. The area considered is partitioned into triangular elements, whose size and shape vary according to both geography and bathymetry. While overall grid size normally approximates

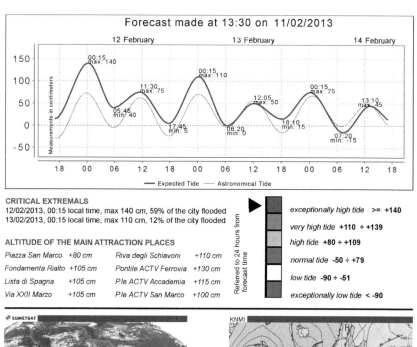

Figure 4.12 Tide forecast released daily by the Tide Center of the Venice Municipality. The bulletin can be downloaded from the Web site[20] of the Municipality of Venice.

35 km in the Mediterranean basin, the subdivision's dimension decreases to 1.5 km in the Adriatic Sea and to approximately 100 m within the Venice Lagoon. Factors forcing water movement include wind speed and barometric pressure, as computed with the aid of the global weather model employed at the European Centre for Medium-Range Weather Forecasts.

Numerous researches and papers have been produced over the last decades on the hydrodynamics of the Venice Lagoon. Hydrodynamic simulations have been used to accurately assess water exchange through the three lagoon inlets in past and present "natural" conditions, and to predict the efficiency of mobile barriers in protecting the city of Venice in years to come. In conjunction with stability and sediment transport equations, hydrodynamic models have also been applied to investigate the major causes driving the long-term evolution of the lagoon since the beginning of the previous century, to detect possible stable configurations of typical lagoon environments, and to forecast trends over the coming decades based on the expected rise in relative sea level. Moreover, lagoon hydrodynamics, alongside the hydrologic response of the ~2000 km^2 lagoon watershed, have been analyzed to assess the possible impact that a given volume of freshwater discharged into the lagoon may have on tidal levels.

The most obvious concern is whether such a large number of MoSE closures, occurring mainly in autumn and winter, might so reduce the dimensions of the lagoon as to turn it into a pond in the near future! The problem is that frequent inlet closures would not be capable of efficiently protecting the lagoon environment, should the number of high tides increase significantly in the years to come. This rise will occur gradually; in any event, as soon as more accurate, reliable modeling predictions of the future sea level rise become available, complementary intervention measures will need to be considered and planned, well in advance.

End Notes

1. http://www.magisacque.it/
2. http://onlinelibrary.wiley.com/doi/10.1002/0470848944.hsa164b/abstract
3. http://wwwrcamnl.wr.usgs.gov/rgws/Unesco/PDF-Chapters/Chapter9-8.pdf
4. http://www.sciencedirect.com/science/article/pii/S0376736106800531
5. http://www.springerlink.com/content/ge4wanlgxaab31v4/
6. http://iahs.info/redbooks/a200/iahs_200_0071.pdf
7. http://www.agu.org/pubs/crossref/1974/WR010i003p00563.shtml
8. http://www.geo.uzh.ch/microsite/rsl-documents/research/SARlab/GMTILiterature/PDF/BH98b.pdf
9. http://www.comune.venezia.it/flex/cm/pages/ServeBLOB.php/L/EN/IDPagina/1748?5079c6ca9f64d

10. http://www.comune.venezia.it/flex/cm/pages/ServeBLOB.php/L/EN/IDPagina/25419
11. http://www.agu.org/pubs/crossref/2009/2008JF001157.shtml
12. http://www.salve.it/uk/soluzioni/acque/mose_iter.htm
13. http://www.salve.it/uk/soluzioni/acque/mose_alternative.htm
14. http://www.salve.it/uk/soluzioni/acque/mose.htm
15. http://www.youtube.com/watch?v=CQQwfiACtzo&feature=player_detailpage
16. http://wwwrcamnl.wr.usgs.gov/rgws/Unesco/PDF-Chapters/Chapter9-3.pdf
17. http://www.sciencedirect.com/science/article/pii/S1474706510000082
18. http://www.salve.it/uk/soluzioni/acque/mose_opere.htm
19. http://www.springerlink.com/content/f121075417614286/
20. http://www.comune.venezia.it/flex/cm/pages/ServeBLOB.php/L/EN/IDPagina/16881

5 Anthropogenic Uplift of Venice by Using Seawater

5.1 The Idea's Origin

It is widely recognized that fluid removal from subsurface reservoirs, in the form of gas, oil, groundwater, geothermal water, and brine, produces a compaction of the depleted formations which migrates totally or partially to the ground surface, thus inducing anthropogenic land subsidence. Its magnitude, time of occurrence, and extent of the area involved depend on a large number of factors, including the amount of fluid withdrawn, the decline in pore pressure, the depth, volume, and permeability of the pumped formation, and the geomechanical properties of the reservoir and overlying porous medium. As mentioned earlier, Venice herself has experienced land subsidence due to groundwater withdrawal in the nearby industrial center of Porto Marghera, mainly during the 1950s and 1960s.

Can the reverse process—the injection of fluids underground—produce the opposite effect at ground surface, namely, land uplift? And can an appreciable uplift reasonably be expected? The answer to these questions is by all means positive from a strictly theoretical viewpoint. The total geostatic load acting upon an aquifer—i.e., the load of the deposits above an underground permeable formation that can transmit a large quantity of water—is counterbalanced by the fluid pore pressure and the stress exchanged by the grain-to-grain contacts: so-called "effective stress." When fluid is withdrawn from a porous medium, pore pressure declines, and the fluid can no longer support as large a percentage of the overlying formations' load. Therefore, more of this load must now be borne by the grain-to-grain contacts of the geological material constituting the matrix of the aquifer system itself, entailing a stress transfer from the fluid to the matrix or granular skeleton of the aquifer system. The increase in effective stress causes a compaction of the subsurface layers. This compaction extends its effect to the ground surface, which therefore subsides.

Conversely, when fluid is injected, pore pressure increases, effective stress consequently decreases, and the injected aquifer expands. The deep expansion migrates totally or partially upward, causing a surface uplift. Generally speaking, the same model, or "equation," governs the two processes, and the resultant subsidence or uplift depends on the sign (negative or positive) of the pore pressure change.

However, land uplift due to subsurface fluid injection has been a much less observed and recognized event, although the practice of pumping fluids into the subsurface principally, as a means of fluid-waste disposal, is more than a half century old. Injection technology has been advancing continuously since its widespread use began in the 1950s–1960s in order to reinject formation water extracted along with hydrocarbons, or to dispose of industrial waste. The injection of water-based solutions, hydrocarbons, CO_2, or N_2 to enhance oil production (enhanced oil recovery, EOR) started in the 1940s and soon became an accepted technique for retrieving additional oil from reservoirs. The earliest measurement of land subsidence due to fluid removal involved the oil field of Goose Creek, San Jacinto Bay, Texas, and dates back to the mid-1920s (Figure 5.1A). However, only in the late 1970s did clear evidence appear of land uplift caused by subsurface fluid injection; this emerged at Long Beach, California, following a massive water injection program implemented to mitigate land settlement due to oil production from the Wilmington oil field (Figure 5.1B).

For a number of reasons, land motion related to subsurface fluid injection went unnoticed for a long time. First, in most cases the disposal of fluids occurred in deserted or sparsely inhabited areas where measuring surface displacements was not a priority, partly given the high costs of traditional leveling surveys. In other instances, uplift was so slight that no environmental hazard was generated, and therefore no monitoring program was really needed; or else the area involved was quite limited, with no reported or foreseen damage to technical structures and infrastructures. In recent times, satellite technology has offered a relatively inexpensive, spatially widespread, accurate methodology for detecting ground movements around the globe, alerting us to any process of anthropogenic uplift wherever it may prove of interest in terms of magnitude, size of the area involved, and time of occurrence. InSAR techniques have developed to an extraordinary degree over the last decade, immensely facilitating the detection and measurement of rising areas in connection with programs of aquifer storage and recharge, injection of water-based solutions and vapors to enhance oil production, CO_2 sequestration in depleted gas fields and/ or saline aquifers, underground CH_4 storage, land subsidence mitigation, and geomechanical characterization of the geologic formations.

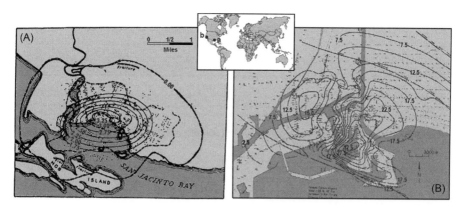

Figure 5.1 (A) Land subsidence (feet) from 1917 to 1926 due to the development of Goose Creek oil field, Texas. (B) Land uplift (cm) at Long Beach, California, from 1958 to 1975 as a result of elastic unloading caused by water injection in the Wilmington oil field. The site locations are shown in the inset.

Several well-documented examples suggest that, although usually a by-product of fluid injection, anthropogenic land uplift occurs worldwide (Figure 5.2). The observed uplift may vary from a few millimeters to tens of centimeters (Figure 5.3B) over a time interval ranging from several months to several years, according to location. The greatest uplifts depend on a number of factors, including the increase in fluid pore pressure, the depth, thickness and extent of the pressurized geological formation, and the hydro-geomechanical properties of the porous medium involved in the process.

Two important examples are shown in Figure 5.3. Since 1988, the Las Vegas Valley Water District, in Nevada, has implemented a groundwater recharge program in the Las Vegas Valley to augment local water supply during periods of high demand. Water is recharged primarily in the coolest months, using treated surface water imported from nearby Lake Mead, on the Colorado River. Since the beginning of the artificial recharge program, groundwater levels have stabilized and recovered: from 1990 to 2005, by as much as 30 m. InSAR measurements for the periods 1995–2000 and 2003–2005 show a general reduction of the subsidence rate, and a broad area of uplift, up to 10 mm/year, adjacent to the easternmost margin of the main artificial recharge zone (Figure 5.3A). The Krechba field is a gas reservoir located in the Algerian Sahara

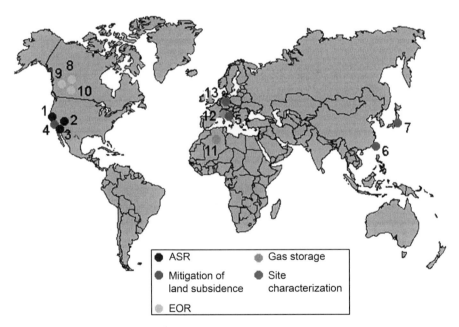

Figure 5.2 Case studies presented in the scientific literature where land uplift due to fluid injection into the subsurface has been observed and measured. The sites are distinguished on the basis of injection purpose: (1) Santa Clara Valley, California, (2) Las Vegas Valley, Nevada, (3) Santa Ana Basin, California, (4) Long Beach, California, (5) Angela-Angelina, Italy, (6)Taipei, Taiwan, (7) Tokyo, Japan, (8) MacKay River, Canada, (9) Pace River, Canada, (10) Cold Lake, Canada, (11) Krechba field, Algeria, (12) Lombardia-1 field, Italy, and (13) Upper Palatinate, Germany. ASR, aquifer storage recharge; EOR, enhanced oil recovery. (For interpretation of the references to color in this figure legend, the reader is referred to the web version of this book.)

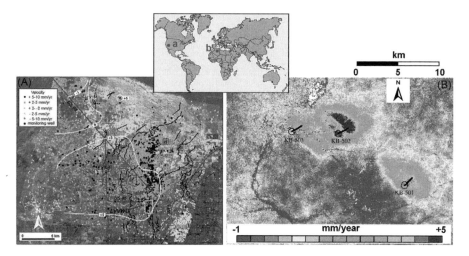

Figure 5.3 Maps of land vertical movement (mm/year) as detected with the aid of InSAR analysis (A) for the northwest portion of Las Vegas Valley, Nevada, between April 2003 and May 2005 and (B) above the Krechba field, Algeria, in the period between 2003 and 2007. *Source*: Modified after Teatini et al. (2011a).[3] (For interpretation of the references to color in this figure legend, the reader is referred to the web version of this book.)

desert at a burial depth of about 2000 m. The gas produced from the Krechba field contains a large proportion of CO_2, which must be separated and disposed of underground for both economic and environmental reasons. The CO_2 is reinjected into the reservoir through three closed wells. Figure 5.3B shows the mean uplift accrued from 2003 to 2007, according to InSAR data. The blue areas are rising by more than 5 mm/year, and are clearly correlated with the location of the three injection wells. The cumulative uplift measured above the Krechba field amounts to about 2 cm.

Recent measurements carried out in the Po Plain sedimentary basin, where Venice is located, also support the likelihood of Venice's rise due to fluid injection into the subsurface. The radioactive marker technique (RMT) is a specific method developed in the field of oil production to measure any deep *in situ* deformation of producing reservoirs. Since the mid-1990s, RMT has been applied in a number of boreholes drilled in the northern Adriatic Sea, spanning a depth of 970–3730 m below msl.[1] After the shutdown of producing wells at the end of their production life, pore pressure within the fields generally rises due to the water inflow from active aquifers surrounding the reservoirs. The ensuing expansion of deep formations has been recorded by RMT. To cope with the energy demand in the cold season, gas and oil companies inject methane gas into exhausted reservoirs during summer and withdraw it during winter. In Italy, Stogit S.p.A. manages a number of 1200-m-deep depleted gas fields for storage purposes in the Po River plain, approximately 150–200 km west of Venice. InSAR data show cyclic uplift and subsidence correlated with the pressure variation within the reservoir. The ground surface above the reservoirs rises every spring–summer by about 10 mm, when reservoir pore pressure increases by 30 bar.[2]

All these well-documented measurements and interpretations have contributed in fostering the concept, enhancing the likelihood, and supporting the hope that similar processes could effectively increase the elevation of Venice relative to msl, and thus help protect the city from *acqua alta*.

5.2 The Complete Project and Pilot Projects for Venice's Uplift

5.2.1 Prediction of Anthropogenic Uplift of Venice with Completion of the Project

A realistic prediction of the possible anthropogenic uplift of Venice has been made using advanced numerical models of the flow-dynamic behavior of injected fluids and the geomechanical response of sedimentary deposits. Several conditions must be observed: (i) injection must take place in salty aquifers to prevent the freshwater resources of the upper 500-m-thick multiaquifer from being contaminated; (ii) injection overpressure must remain within a safety limit—in our case estimated at 20 bar—in order to preserve the integrity of the injected formations and the overlying ones: the aim is to avoid causing fractures that might yield an abnormal dissipation of overpressure, and jeopardize the sealing capability of the caprock confining the injected sequences; (iii) the uplift of Venice must be as uniform as possible, without giving rise to differential displacements that might threaten the safety, stability, and integrity of the city's historical buildings and monuments.

The modeling approach comprises several steps that are summarized as follows: (1) selection of the simulation domain, i.e., the portion of the underground porous medium where the physical process occurs; (2) definition of the most representative geological properties, such as the hydraulic conductivity and geomechanical compressibility of the rock; (3) model calibration and validation, i.e., attuning the input parameters to match the observed records; and (4) use of the model in its predictive capacity, i.e., computation of the system's response to water injection.

In our study, the simulation domain is a box 50×60 km comprising the entire lagoon and centered on Venice. The top surface is defined by a digital elevation model of the area, and the bottom surface is set at a depth of 10 km. Such fairly large dimensions are required to ensure a negligible effect of the boundary conditions on the results we are interested in. For the outcome to be reliable, the model needs a good input dataset, as is provided by more than 30 years of hydro-geomechanical research along the northern Adriatic coastland. A very accurate 3D reconstruction of the Quaternary geology has been made from approximately 1050 km of multichannel seismic profiles, eight exploration wells (Figure 5.4), and several shallow wells drilled for purposes of groundwater pumping throughout the area. The seismic data have allowed us to define the most important geologic formations at unprecedented detail, down to a depth of 2000 m. Five main geologic sequences, labeled from bottom to top as PLS1 through PLS5, have been identified within the Pleistocene, and two (PLC1 and PLC2) within the Pliocene (Figure 5.5). Note the great geometric variability of these units,

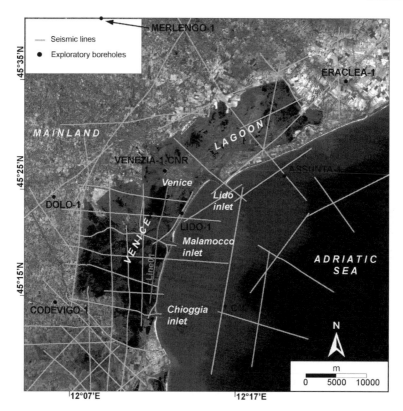

Figure 5.4 Seismic lines and exploratory boreholes used to reconstruct the 3D geologic setting of the Venetian subsurface basin.[4] (For interpretation of the references to color in this figure legend, the reader is referred to the web version of this book.)
Source: Modified after Teatini et al. (2011b).

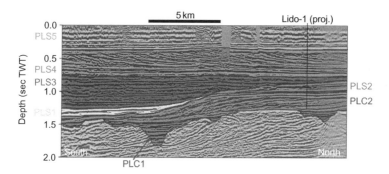

Figure 5.5 Seismic sequences from the profile Line01 highlighted in blue in Figure 5.4. The geological sequences are identified by different colors. (For interpretation of the references to color in this figure legend, the reader is referred to the web version of this book.)
Source: Modified after Tosi et al. (2012).[5] (For interpretation of the references to color in this figure legend, the reader is referred to the web version of this book.)

alternatively originated by deposition of Alpine and Apennine sediments; pinch out one against the other, and significantly change their thickness in the north-to-south direction. The seismic survey provides no clear evidence of major faults below the Venice coastland, as far down as the depth involved in the project. The geologic sequences have been integrated with the geophysical logs from the available deep boreholes, thus providing a hydrogeologic characterization of the mapped formations.

The units of potential interest for injection are PLS2, PLS3, and PLC2, spanning a depth of range 650–1000 m. They appear to be well confined at the top by the continuous low-permeability sequence, PLS4, which represents an important barrier against any upward migration of the injected fluid (Figure 5.5).

Various litho-stratigraphic layers have been identified within the proposed injection sequences and the overlying and underlying sedimentary deposits. These include sand, silty sand, silt, and clay with a hydraulic conductivity K that ranges over five orders of magnitude, i.e., 10^{-5} to 10^{-10} sm/s, as measured by the Italian National Research Council using samples from borehole Venezia-1-CNR drilled at Tronchetto, Venice, and by Eni E&P, the Italian national oil company, on sands and clays from on- and off-shore exploratory wellbores along the northern Adriatic coastland (Figure 5.4). The geomechanical properties, in particular vertical rock compressibility c_M, are of paramount importance for a reliable prediction of Venice's uplift. Based on deep RMT measurements and shallower laboratory tests, a constitutive law has been derived, with c_M defined as a function of depth and vertical effective stress. For sequences PLS2, PLS3, and PLC2, c_M varies between 3×10^{-4} and 1×10^{-4} bar^{-1}. These values refer to rock compressibility in elastic virgin loading (cycle I) conditions, i.e., during compaction. However, injection involves the unloading of sediments (cycle II), i.e., rock expansion. Generally, sedimentary deposits present significant mechanical hysteresis, with c_M greater in loading than in unloading conditions. *In situ* RMT measurements in the northern Adriatic basin show that $s = c_{M,\text{loading}}/c_{M,\text{unloading}}$ decreases from 3 to 1.5 in the depth range of 2700–7000 m. Laboratory tests by Eni E&P suggest s values around 3 at a 1000-m depth. Finally, recent modeling studies of seasonal land motion above gas storage reservoirs in the Po River basin indicate an s value between 3.5 and 4 for 1000–1200 m depth. Thus, a value of 3.5 is used in the modeling prediction described below.

Figure 5.6A shows a plan view of the 3D tetrahedral finite element (FE) mesh of the litho-stratigraphy below the lagoon. Figure 5.6B provides an axonometric representation of the 3D grid along a cross section parallel to the northern Adriatic coastline, and highlights the complexity of the geologic structure. The high-resolution mesh consists of 1,905,058 tetrahedrons and 328,215 nodes, and accurately reproduces the heterogeneous litho-stratigraphic sequence (Figure 5.7). The mesh is greatly refined close to Venice, where the injection wells are planned.

The injection would occur through a curtain of 12 wells located on a 10-km-diameter circle centered on the Rialto Bridge in Venice (Figure 5.6A). The well layout has been planned in such a way as to guarantee a uniform pattern for the overpressure field, and hence the uplift, in correspondence to the historical city.

The prediction of pore overpressure and the resulting anthropogenic uplift results from a numerical solution of the classical saturated flow and poroelastic equations. The time behavior of the injection rate and the related pore pressure at each borehole

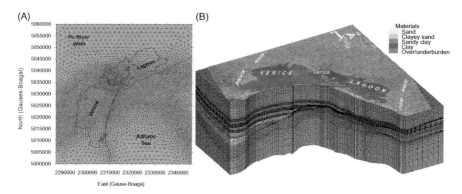

Figure 5.6 (A) Plan view of the 3D FE grid. The location of the injection wells is shown. (B) Axonometric view of the 3D FE grid sectioned along the coastline. The colors are representative of the various lithotypes detected within the PLS3, PLS2, and PLC2 formations and the overlying and underlying units. The vertical exaggeration is 5. (For interpretation of the references to color in this figure legend, the reader is referred to the web version of this book.)

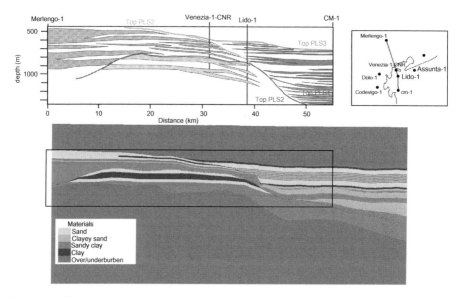

Figure 5.7 Hydrogeologic north-south cross section of the Venetian basin as obtained from the combined use of seismic data and well logs (above) and as reconstructed in the 3D FE model (below). The tops of formations PLS3, PLS2, PLS1, and PLC2 are highlighted.
Source: Modified after Teatini et al. (2011b). (For interpretation of the references to color in this figure legend, the reader is referred to the web version of this book.)

are accurately computed with the aid of the so-called "well model." The simulations were performed using codes FLOW3D and GEPS3D, developed at the University of Padua.

The planned uplift is expected to occur after 10 years, i.e., a period comparable to the time originally planned for MoSE's construction. The injection would take place

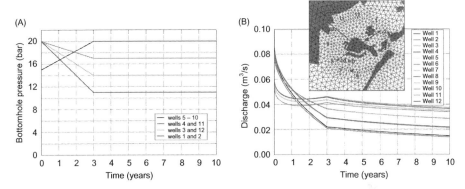

Figure 5.8 (A) Well bottom overpressure and (B) injection rate versus time according to the borehole numeration provided in the inset. (For interpretation of the references to color in this figure legend, the reader is referred to the web version of this book.)
Source: Modified after Teatini et al. (2011b).

within the sandy layers of formations PLS3 (video1), PLS2 (video2), and PLC2 (video3). For video files, please refer http://booksite.elsevier.com/9780124201446

Owing to the lateral variability of the geologic sequences, not all the pumping wells necessarily intersect the three units (video4). To abate differential displacements, the pore overpressure in each single borehole and the corresponding injection rate are attuned as shown in Figure 5.8A and B, respectively. The injection rate ranges from 10^{-2} to 10^{-1} m³/s per well, with a total injected volume of 135 Mm³ over 10 years. The injected seawater is to be chemically treated for compatibility with the water of the injection formation. The modeling results are very encouraging. Figure 5.9 provides the predicted anthropogenic uplift of Venice after 1, 2, 5, and 10 years of injection, and shows an ultimate, fairly uniform surface uplift of about 26 cm for the whole city. Uplift behavior over the simulated time interval at the world-famous Rialto Bridge, in the center of Venice, is shown in Figure 5.10. The maximum rate of 5 cm/year occurs between the second and the third years, and decreases thereafter. It is worth noting that a 26-cm uplift is a remarkable value in relation to the Venetian setting.

Indeed, such an uplift would have considerably mitigated the *acqua alta* recorded in the city from 1872 to the present (Figure 4.2), canceling 90% of the events higher than 110 cm, and greatly attenuating the remaining 10%.

Considering predicted increases in eustatic sea level accompanying global climate change expected to take place in the next few decades, and the natural land settlement of the lagoon, a ~26 cm uplift might offset the relative sea level rise of the northern Adriatic Sea as predicted by the end of the present century (Figure 4.11), greatly decreasing the frequency of MoSE's activation, extending its useful life, and reducing the environmental impact of a restricted water exchange between the lagoon and the sea.

5.2.2 Testing the Full-Scale Injection Project by a Pilot Project

In view of the precarious lagoon environment and the great artistic heritage of the city of Venice, anthropogenic uplift by seawater injection would be inconceivable

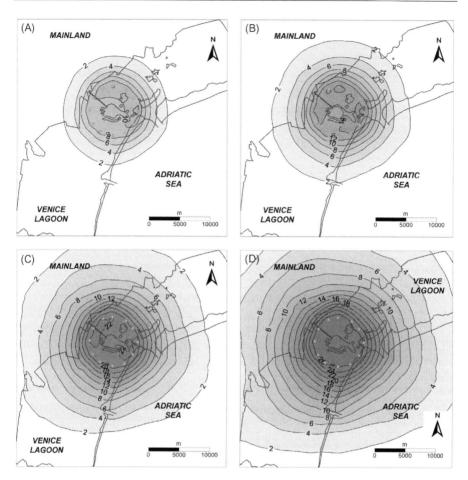

Figure 5.9 Predicted uplift (cm) after (A) 1, (B) 2, (C) 5, and (D) 10 years of injection. The injection wells are marked in green. (For interpretation of the references to color in this figure legend, the reader is referred to the web version of this book.)
Source: Modified after Teatini et al. (2011b). (video5).

without extremely detailed knowledge of the underground system and a reliable prediction of actual subsurface response to planned pumping. We can attain such in-depth knowledge through the layout and implementation of a preliminary pilot project entailing new geophysical investigations and an *ad hoc* injection experiment.

The purposes of a pilot project are manifold. The first is to calibrate and finely attune the hydrological and geomechanical models in order to improve their predictive capacity, thus constructing a fundamental tool for forecasting ground deformation and for keeping the injection process under careful control. The second is to measure the resulting land motion using the most advanced satellite technology, showing that the joint use of real-time measurements and numerical predictions can

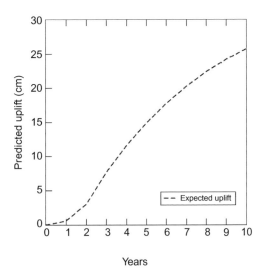

Figure 5.10 Predicted uplift versus time at Rialto Bridge.

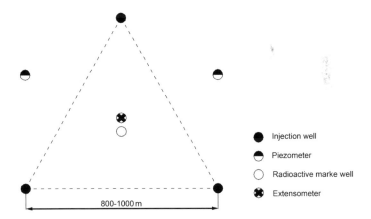

Figure 5.11 Pilot project: layout of the injection wells and other instrumented boreholes. *Source*: Taken from Castelletto et al. (2008).[6]

provide an effective methodology for monitoring and controlling overall occurrence, in a prospective future program for the anthropogenic uplift of Venice. And last but not least, we aim to improve the geological and litho-stratigraphic representation of the lagoon subsurface down to the depth involved.

Figure 5.11 shows a preliminary design of the pilot project components. Three wells from which saltwater is to be injected underground are located at the vertices of a regular triangle with sides 800–1000 m long, and drilled down to a depth of approximately 1000 m. Each single well will inject at a rate required to safely uplift the study area by a few centimeters, i.e., an amount that can be accurately monitored during the 3-year duration of the experiment. The main quantities to be measured during the injection process include pore water overpressure in both the injection wells and at least two piezometers; the injection rate from each individual well; expansion of the injected formations and overlying clay layer, monitored using RMT; compaction, if any, of the upper freshwater aquifer system, using an extensometer device; and land surface displacements—horizontal and vertical—monitored using remote sensing technologies (GPS and InSAR techniques) and more traditional high-precision spirit leveling.

A fundamental step will be to select the site, based on both geologic and logistic considerations. From the geologic point of view, the aquifer where saltwater is injected should possess geometric and lithological features similar to those present in the full project. Logistically speaking, the selected area must fulfill a number of safety requirements, e.g., its ground elevation should be sufficient to avoid any risk of flooding due to likely events of *acqua alta* during the experiment, and sufficiently distant from vulnerable sites—such as factories, villages, or sensitive areas—that might be damaged should the experiment evolve differently than planned. Moreover, the site must offer relatively easy, inexpensive accessibility by trucks in order to install and remove the experimental equipment required by the injection plant. Available amounts of water for injection should also be ensured, at a relatively short distance from the lagoon, with easy access to an *ad hoc* treatment plant for geochemical compatibility.

At this preliminary stage, four possible choices would seem to offer a balanced approach to the above requirements, and comply with the principal geologic and logistic constraints (Figure 5.12). Le Vignole Island is probably one of the most interesting sites from a geologic viewpoint, given its proximity to Venice, but for purposes of this pilot study the island can be accessed only by ship. The site of San Giuliano, on the other hand, whose facilities would make truck access and the installation of experimental equipment quite convenient, is so near such crucial infrastructures as the major motorway overpass that the injection experiment could potentially threaten the safety of those structures. Other sites of interest are Fusina and Cascina Giare, the latter of the two located farther from Venice; both are easily accessible by truck, and appear to be far enough from vulnerable areas.

The experiment has been simulated numerically using a model of subsurface water flow and land uplift similar to the one implemented for the full project, plus a representative schematization of the geology involved. The simulation employs a constant rate equal to $0.012\,m^3/s$ of saltwater, i.e., an amount significantly smaller than amounts foreseen for the completed project (Figure 5.8B), planned for continuous injection into each well. For example, Figure 5.13 shows the outcome for the Fusina simulation in terms of water overpressure and land uplift at the end of the 3-year experiment. Greatest overpressure and uplift—6.7 bar and 7.3 cm, respectively—are expected to occur at the center of the imaginary triangle. The considerably limited amount of predicted pore overpressure should prevent any risk of creating hydraulic fractures. Note the uniform spatial distribution of the uplift, consistent with the fairly regular distribution of overpressure. Figure 5.14 shows the time-dependent responses during and after the experiment, consistent with the cessation of injection on completion of the third year. It is worth noting that,

Anthropogenic Uplift of Venice by Using Seawater

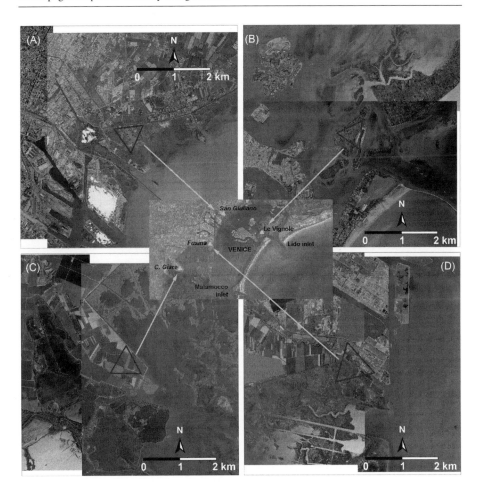

Figure 5.12 Aerial view of the area of (A) San Giuliano, (B) Le Vignole, (C), Cascina Giare, and (D) Fusina. Their location is shown in the central inset. The blue triangle indicates the preliminary trace of the margin of the experimental site. (For interpretation of the references to color in this figure legend, the reader is referred to the web version of this book.)
Source: Modified after Castelletto et al. (2008).

should injection continue at a lower rate after the third year, ground level could be kept constantly stable at the elevation recorded at the end of the trial.

5.3 Safety and Stability of Venice's Uplift

A major concern regarding the full-scale project is the possible creation of a nonuniform uplift that might jeopardize the stability and integrity of the city's historical buildings and monuments. In civil engineering, this event is framed as "differential

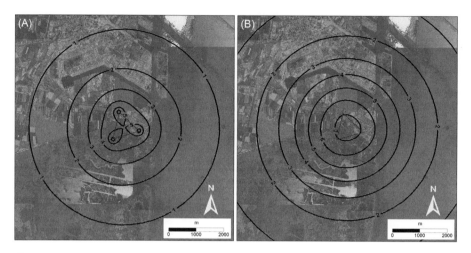

Figure 5.13 (A) Pore water overpressure (bar) averaged over the injected aquifer thickness, and (B) land uplift (cm) at the completion of the pilot project (Fusina site).
Source: Modified after Castelletto et al. (2008).

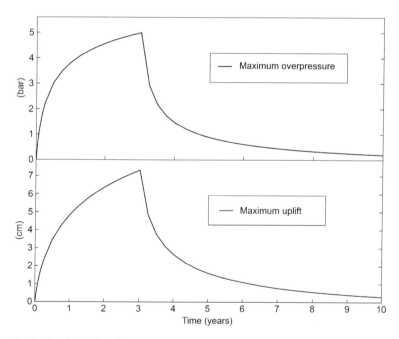

Figure 5.14 Time behavior of pore water overpressure (bar) averaged over the injected aquifer thickness, and land uplift (cm) at the center of the ideal injection triangle.
Source: Modified after Castelletto et al. (2008).

displacement," i.e., the displacement of two random locations in the city relative to the distance separating them. The value attached to this factor is quite variable; basically, it hinges on the perception of acceptable damage to any specific manmade structure. Generally speaking, a rather limited differential displacement must be established wherever a criterion of an aesthetic or functional nature prevails, e.g., for an important historical construction, or a civil structure, such as a bridge or dam. A less binding limit might be accepted, instead, for certain industrial buildings.

We have taken into consideration three basic thresholds for differential displacement. The first value is strictly related to the peculiarity of Venetian buildings, most of which are made of ordinary brick and wood. The most severe restriction acceptable for multistory masonry buildings recommends not exceeding 50×10^{-5}, i.e., a vertical displacement difference of no more than 50 mm per 100 m. The other two limits refer to differential displacements already experienced by Venice in the past, with no damage reported to the existing architectural patrimony. As mentioned above, the historical center of Venice was seriously affected by anthropogenic land subsidence during the 1950s and 1960s as a result of groundwater overdraft in the nearby industrial mainland. The processing of leveling data has allowed experts to map the differential displacements occurring in the city during that period, with values of up to 10×10^{-5} resulting for some areas (Figure 5.15A). Finally, a recent investigation of vertical displacements in Venice has been performed using InSAR, processing a series of 30 TerraSAR-X images acquired between March 5, 2008 and January 29, 2009. The differential displacements resulting from these high-resolution measurements clearly show a substantial stability of the city as a whole. However, they also reveal that important differential displacements, in the order of 100×10^{-5}, are presently occurring around a few adjacent buildings, i.e., over a few tens of meters, most likely due to surface loads, heterogeneities of the upper Holocene deposits upon which the city is founded, and restoration works on canals and *fondamenta* facing building blocks. An example is shown in Figure 5.15B

With these premises in mind, we may look at Figure 5.16, showing the expected distribution of differential displacements across the lagoon as predicted for the most critical year, the one following the 10-year injection period (Figure 5.9D). A perusal of this figure suggests that the largest differential displacement throughout the lagoon is less than 5×10^{-5}, i.e., 5 mm per 100 m, decreasing to 0.1×10^{-5}, i.e., 0.1 mm per 100 m, within the area encompassed by the wells: namely, the entire city. The differential displacements resulting from our project of seawater injection are 500 times smaller than the required technical limit, and from 100 to 1000 times smaller than the values experienced by Venice during the past 50 years.

Despite the complex geometry of the injected formations (Figure 5.7), as a major result of the thick overlying formations' attenuating effect, the expected uplift would be highly uniform throughout the city. Therefore, on the basis of the simulated differential displacements, the uplift represents no threat to the monuments' and buildings' safety.

Finally, a further safety issue is worth discussing. Hydrogeology experts widely agree that even within the same geological formation, such as a sandy aquifer, soil permeability may be highly heterogeneous with spatial variability ranging over many orders of magnitude.

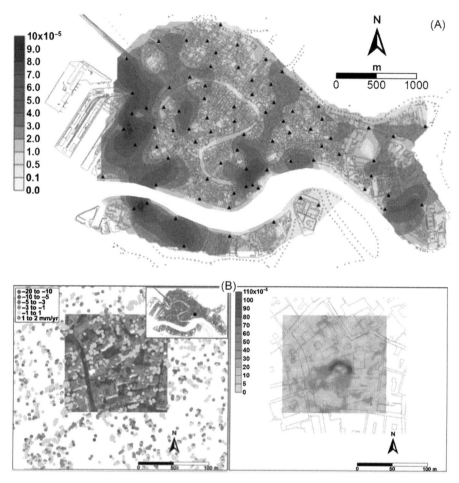

Figure 5.15 (A) Differential displacements at Venice as obtained from high-precision leveling records over the period 1961–1969. (B) Average displacement rates from March 2008 to January 2009 for a portion of Venice measured using InSAR (left) and differential displacements as derived from the displacements (right). (For interpretation of the references to color in this figure legend, the reader is referred to the web version of this book.)
Source: Modified after Gambolati et al. (2009).[7]

Such variability can give rise to a highly irregular overpressure distribution and expansion of the injected formation. The obvious question arising here is whether a heterogeneous permeability might produce an aquifer expansion so irregular that an important fraction of the uneven *in situ* deformation migrates to the ground surface, thus jeopardizing the uniformity of the uplift and endangering Venice's monuments.

To address this issue, a statistical analysis has been applied to the pilot project, starting from the assumption that rock permeability can vary up to four orders of magnitude (e.g., see Figure 5.17) relative to the mean value used to compute the prediction in Figure 5.13.

Anthropogenic Uplift of Venice by Using Seawater

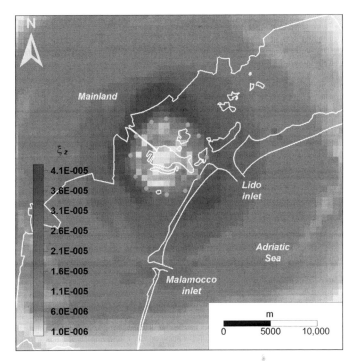

Figure 5.16 Expected differential displacements 10 years after pumping inception. The injection boreholes are marked in green. (For interpretation of the references to color in this figure legend, the reader is referred to the web version of this book.)
Source: Modified after Teatini et al. (2011b).

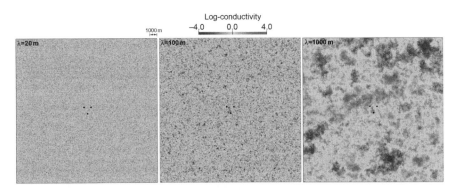

Figure 5.17 Examples of permeability distribution characterized by a correlation length λ equal to 20, 100, and 1000 m. Parameter λ describes the distance beyond which the permeability values of two sites are stochastically independent. Red and blue zones are representative of permeability areas four orders sof magnitude larger and smaller than the average, respectively. The injection wells are shown by black dots. (For interpretation of the references to color in this figure legend, the reader is referred to the web version of this book.)
Source: Modified after Teatini et al. (2010).[8]

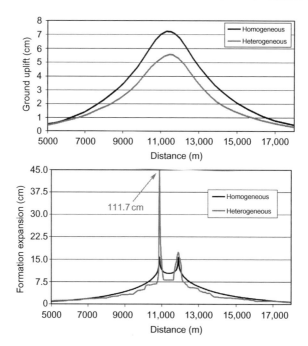

Figure 5.18 Injected formation expansion and land uplift along a vertical cross section through two wells of the pilot experiment. The results obtained with the most pessimistic permeability distribution are compared with the case of a homogeneous aquifer system. *Source*: Modified after Gambolati et al. (2009). (For interpretation of the references to color in this figure legend, the reader is referred to the web version of this book.)

As required by statistics, several spatial patterns of permeability were created, and 1000 simulations were performed using flows and geomechanics consistent with the 1000 permeability scenarios. The most critical solution—i.e., the one with the most irregular overpressure distribution, and hence expansion, of the injected formation—was then extracted. Figure 5.18 shows that for the worst case (i.e., the most irregular aquifer expansion), anthropogenic land uplift appears to be quite smooth, similar to that shown in Figure 5.13B. This provides evidence of the important smoothing effect exerted by the sedimentary deposits overlying the pumped aquifer: an effect which helps mitigate and spread the local deformation at the land surface.

5.4 Can Venice Be Raised Economically, in an Environmentally Friendly Way?

Predictably, costs for the implementation and maintenance of the proposed injection system should not add significantly to the high expenditures required for the protection and safeguarding of Venice. In 2008 a fairly accurate cost estimate was made for the pilot project, i.e., three injection boreholes, assuming a 20-bar well overpressure and a $0.02\,m^3/s$ pumping rate per well, quite consistent with the complete project design. Note that, since the technology used is oil field technology, the budget may vary significantly even over a short term ranging from 6 to 12 months. Costs can be subdivided into two categories: construction costs and annual operation costs. As for construction, the major costs involve borehole drilling (2.5 M€ for each well), the plant for seawater intake (0.5 M€), filtering (1.3 M€), chemical treatment (0.3 M€), and pumps (0.5 M€ each), plus a negligible cost for pipes. The overall budget amounts to 11.1 M€. Annual operation cost may be estimated at 1.4 M€/year and involves electricity (0.5 M€/year), chemicals (0.6 M€/year), maintenance (0.2 M€/year), and personnel (0.1 M€/year). We can make a reasonable estimate of the total construction and annual operation costs for the full-scale project if we multiply the above numbers by a factor ranging from 5 to 10, 7.5 on average. Rounded off, the total sums are 80 M€ and 10 M€/year, respectively.

Is this too much or too little? To answer this question, compare the corresponding expenditure for MoSE, whose construction is currently expected to cost over 5000 M€ (see Section 3.3) and whose management and maintenance—according to the latest estimate— amount to 30 M€/year. Monitoring will cost another substantial sum. An official document issued by the Venice Water Authority estimates that the annual cost required for research and monitoring throughout the lagoon will be about 17 M€/year. In this perspective, the expenditure for monitoring the uplift project seems quite modest. After an initial cost of approximately 2 M€ for the radioactive marker borehole and the extensometer implementation, plus 4 M€ for placing a pair of boreholes for fluid pressure measurements, the annual cost for measuring land displacement using SAR-based techniques and permanent GPS stations amounts to only 0.5 M€/year. The costs are, of course, comparable in the pilot and full-scale projects.

The final point we wish to discuss concerns the injection's possible impact on the lagoon environment. Generally speaking, the infrastructures required to implement the uplift project appear to be environmentally safe, with little or no foreseeable impact on the lagoon ecosystem. But this is not all. As noted previously, since the early 1900s, Venice's lagoon has been gradually evolving. Its tidal flats have deepened, its salt marshes diminished in number. These processes stem from a number of factors, including both natural and anthropogenic land subsidence, and the rise in msl owing to global climate changes. In this respect, the uplift of the lagoon floor, a by-product of our main objective of raising Venice's historical center, might well exert a positive influence on the morphology of this constantly fluctuating environment, since it may help to reduce the negative budget, i.e., to offset the loss of sediments exchanged yearly through the inlets, between the lagoon and the Adriatic Sea.

The Venice Water Authority has recently suggested that anthropogenic uplift be investigated as an experimental methodology aimed at mitigating the present morphological deterioration of the lagoon.

We have examined the influence of predicted uplift on the lagoon bathymetry, using as a reference the bathymetry surveyed in 2002 (Figure 5.19A). Figure 5.19B shows the expected new bathymetry after 10 years of seawater injection, based on the uplift displayed in Figure 5.9D.

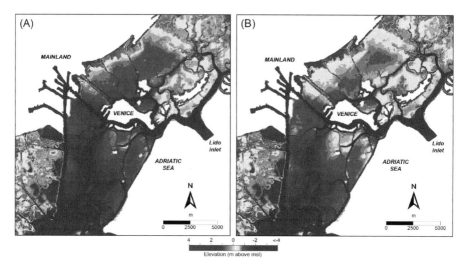

Figure 5.19 Bathymetry of the central part of the Venice lagoon (A) as of 2002 and (B) as modified at the end of the complete injection project. (For interpretation of the references to color in this figure legend, the reader is referred to the web version of this book.).
Source: Taken from Teatini et al. (2011b).

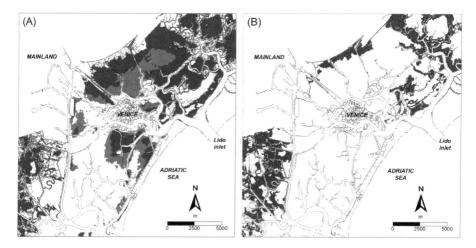

Figure 5.20 Changes (in red) (A) of the lagoon bottom with elevation between − 1 and 0 m above msl and (B) of the emerged areas with elevation between 0 and 1 m above msl of the 2002 lagoon bathymetry (in blue) as predicted at the end of complete injection project.
Source: Taken from Teatini et al. (2011b). (For interpretation of the references to color in this figure legend, the reader is referred to the web version of this book.)

The bathymetry's outcome depends not only on the uplift itself, but also on the depth of the affected sea floor. In fact, the lagoon environment encompasses high-value morphological features such as salt marshes, which lie at an average elevation higher than msl—usually between 0.1 and 0.5 m—and shallows, i.e., flat areas just below msl, approximately 0.0–1.0 m. A depth change from 1 cm to 1 dm has a much stronger effect on these areas than on tidal flats, which lie approximately between 1.0 and 2.0 m, or tidal channels, which in the Venetian Lagoon usually present a water level ranging from 2 to 3 m. Figure 5.20 shows the differences between present and expected bathymetry for areas where the lagoon bottom is between − 1.0 and 0.0 m (Figure 5.20A), and for areas currently above msl (Figure 5.20B). The emerging areas appear to expand by very little, while the shallows widen quite significantly around the city.

End Notes

1. http://www.onepetro.org/mslib/app/Preview.do?paperNumber=00079470&societyCode=SPE
2. http://www.agu.org/pubs/crossref/2011/2010JF001793.shtml
3. http://www.sciencedirect.com/science/article/pii/S0264370710000931
4. http://www.agu.org/pubs/crossref/2011/2011WR010900.shtml
5. http://jgs.geoscienceworld.org/content/169/5/507.abstract
6. http://www.agu.org/pubs/crossref/2008/2007WR006177.shtml
7. http://onlinelibrary.wiley.com/doi/10.1111/j.1365-3121.2009.00903.x/abstract
8. http://www.agu.org/pubs/crossref/2010/2010WR009161.shtml

Conclusion

In the past, Venice's relationship with the waters of her surrounding lagoon drastically conditioned her political, economic, and military history. The key to Venice's wealth and power has always been her delicate and precarious lagoon environment, which has survived practically unchanged for nearly a millennium, thanks to her farsighted inhabitants.

The city's economic decline, which began in the Renaissance, is still proceeding today in nearly every sector except for tourism, which increases every year.

Over the past few centuries Venice has been steadily losing elevation relative to sea level, creating the premise for her very disappearance in a not-too-distant future. Her demise would be a disaster both for the human spirit and for a tourist-based economy. Can Venice rise above the water now jeopardizing her existence?

We strive to find an answer. The proposed subsurface injection project may yield a viable solution by offering a technologically simple, relatively inexpensive, and environmentally friendly means for protecting Venice from the sea's attack. *Acqua alta* is the most dangerous threat to Venice's priceless artistic and cultural heritage and—indeed—to her very survival. If our project is carried to completion, it may raise Venice by 25–30 cm.

The project consists of injecting seawater into a number of sandy layers interspersed within a brackish water aquifer system lying between 650 and 1000 m below the surface of the lagoon.

After being chemically treated to create compatibility with the waters present in this microenvironment, seawater would be injected into the target geologic formation for a period of 10 years, through 12 wells encompassing the city. Later on, the wells would remain operative in order to keep the uplift stable, although the injection rate could be considerably reduced. The project would proceed under conditions of absolute safety for the integrity and stability of monuments, palaces, and buildings, because the predicted differential displacements—i.e., the displacements of two random locations in the city relative to the distance separating them—would be much smaller both than those caused in Venice by land subsidence occurring in the 1950s and 1960s, and those presently under way, as accurately revealed by satellite interferometry.

The predicted uplift of 25–30 cm results from a series of simulations performed with the aid of advanced hydrodynamic and geomechanical models. These models are based on the latest acquired *in situ* deformation data for the northern Adriatic, and on studies of land settlement and uplift caused by gas production and gas storage in the Po River plain.

The idea of raising Venice anthropogenically is in harmony with the MoSE system. First of all, the uplift would have practically no measurable impact on the lagoon ecosystem. Second, an uplift could entirely offset a sea level rise due to global warming equal to the uplift itself, therefore ensuring MoSE's functionality throughout its expected 50-year operation lifetime, or even longer, should the sea rise along with the uplift as foreseen by our project. In a word, it might contribute to the success of the mobile gates in prospective as well under new sea level conditions.

The uplift would also mitigate the scourge of extreme and moderate *acqua alta*. Not only would this bring great benefit from the very beginning, but it would also annul any need to activate MoSE more and more frequently in the future, which would entail an increasingly severe impact on the lagoon environment.

The 950 km of seismic lines extended under the lagoon during the 1970s were recently made available by ENI-E&P, the Italian oil company, along with 100 km of new lines laid out in the Adriatic, in front of the Lido. This has permitted construction of a novel geologic and litho-stratigraphic model of the Venetian subsurface, offering an unprecedented level of accuracy and reliability of prediction.

Moreover, an injection rate appropriate for each single well, ranging in space and time from $0.01–0.1\,m^3/s$, would ensure Venice a very even elevation, with differential displacements on the order of 0.5 mm/100 m, i.e., one hundred times below the limits suggested in the technical literature for masonry constructions. This would dissipate worries over risks associated with sea level rise threatening Venice's precious monumental patrimony.

A pilot experiment should be carried out to verify the feasibility of any project for uplifting Venice. The pilot experiment plan foresees three boreholes located at the vertices of a triangle with sides 1 km long, in a lagoon area to be selected (suitable sites might be Cascina Giare, Fusina, Le Vignole, and San Giuliano). The aim would be as follows: (1) to obtain further detailed litho-stratigraphies capable of enhancing our knowledge of the underground lagoon, (2) to perform an injection test with (treated) seawater and measure the overpressure generated in the injected formation, (3) to monitor continuously and in real time the actual land uplift in the area, with the aid of high-precision leveling, GPS, and satellite interferometry, and (4) to set-up and experiment with a procedure of optimal control, for instance the uniformity of uplift may be checked with the aid of a sensor feedback that accommodates automatically the injection rate in each single well.

The pilot experiment is planned to last for 3 years, entailing a maximum expected uplift of 7 cm at the triangle's center. The experiment's results would hopefully confirm the feasibility of safe and effective anthropogenic uplift of Venice; define a more accurate, ultimate geohydrologic and geomechanical model of the lagoon subsurface, to be used in predictions for the full-scale program; and provide the operative details to be implemented in enacting the full-scale project of raising Venice by deep seawater injection.

References

Abbott, A., 2004. Plans resurrected to raise Venice above encroaching sea. Nature 427 (6971), 184.
Ammerman, A.J., McClennen, C.E., 2000. Saving Venice. Science 269, 1301–1302.
Baù, D., Ferronato, M., Gambolati, G., Teatini, P., 2002. Basin-scale compressibility of the Northern Adriatic by the radioactive marker technique. Geotechnique 52 (8), 605–616.
Bell, J.W., Amelung, F., Ferretti, A., Bianchi, M., Novali, F., 2008. Permanent scatterer InSAR reveals seasonal and long-term aquifer system response to groundwater pumping and artificial recharge. Water Resour. Res. 44. http://dx.doi.org/10.1029/2007WR006152.
Bock, Y., Wdowinski, S., Ferretti, A., Novali, F., Fumagalli, A., 2012. Recent subsidence of the Venice Lagoon from continuous GPS and interferometric synthetic aperture radar. Geochemistry Geophysics Geosystems G3 13 (3) 13, pp.
Brambati, A., Carbognin, L., Quaia, L., Teatini, P., Tosi, L., 2003. The Lagoon of Venice: geological setting, evolution and land subsidence. Episodes 26 (3), 264–268.
Bras, R.L., Harleman, D.R.F., Rinaldo, A., Rizzoli, P., 2002. Obsolete? No. Necessary? Yes. The gates will save Venice. EOS Trans. 83 (20), 217–224.
Camuffo, D., Sturaro, G., 2003. Sixty-cm submersion of Venice discovered thanks to Canaletto's paintings. Clim. Change 58, 333–343.
Canestrelli, P., Mandich, M., Pirazzoli, P.A., Tomasin, A., 2001. Wind, Depression and Seiches: Tidal Perturbations in Venice (1950–2000). Centro Previsioni e Segnalazioni Maree, Venice, Italy, 155 pp.
Carbognin, L., Gatto, P., Mozzi, G., Gambolati, G., Freeze, R.A., 1979. The Subsidence of Venice: A Danger Overcome, TR by National Research Council of Italy, 33 pp.
Carbognin, L., Teatini, P., Tosi, L., 2004. Eustacy and land subsidence in the Venice Lagoon at the beginning of the new millennium. J. Mar. Syst. 51 (1–4), 345–353.
Carbognin, L., Teatini, P., Tomasin, A., Tosi, L., 2010. Global change and relative sea level rise at Venice: what impact in term of flooding. Clim. Dyn. 35, 1039–1047.
Carniello, L., Defina, A., Fagherazzi, S., D'Alpaos, L., 2005. A combined wind wave–tidal model for the Venice lagoon, Italy. J. Geophys. Res. 110, F04007. http://dx.doi.org/10.1029/2004JF000232.
Castelletto, N., Ferronato, M., Gambolati, G., Putti, M., Teatini, P., 2008. Can Venice be raised by pumping water underground? A pilot project to help decide. Water Resour. Res. 44, W01408. http://dx.doi.org/10.1029/2007WR006177.
Chin, G., 2004. High and dry. Science 303, 146.
Comerlati, A., Ferronato, M., Gambolati, G., Putti, M., Teatini, P., 2003. Can CO_2 help save Venice from the sea? EOS-Trans. Amer. Geophys. Union 84 (49), 546–553.
Comerlati, A., Ferronato, M., Gambolati, G., Putti, M., Teatini, P., 2004. Saving Venice by sea water. J. Geophys. Res.-Earth Surf. 109 (F3) F03006, http://dx.doi.org/10.1029/2004JF000119.
Comerlati, A., Ferronato, M., Gambolati, G., Putti, M., Teatini, P., 2005. A coupled model of anthropogenic Venice uplift. In: Abousleiman, Y. (Ed.), Poromechanics - Biot Centennial 1905–2005. Taylor & Francis Group Publ., London (UK), pp. 317–321.

Comerlati, A., Ferronato, M., Gambolati, G., Putti, M., Teatini, P., 2006. Fluidynamical and geomechanical effects of CO_2 sequestration below the Venice Lagoon. Environ. Eng. Geosci. 12 (3), 211–226.

D'Alpaos, L., 2010. Fatti e Misfatti di Idraulica Lagunare, Istituto Veneto di Scienze Lettere ed Arti, 329 pp.

Ferronato, M., Gambolati, G., Putti, M., Teatini, P., 2008. A pilot project using seawater to uplift Venice anthropogenically. EOS-Trans. Amer. Geophys. Union 89 (16), 152.

Ferronato, M., Gambolati, G., Teatini, P., 2011. The role of aquifer heterogeneity in the anthropogenic uplift of Venice. In: Wang, Y. (Ed.) Calibration and Reliability in Groundwater Modelling: Managing Groundwater and the Environment, IAHS Publication 341, 252-257.

Gabriel, A.K., Goldstein, R.M., Zebker, H.A., 1989. Mapping small elevation changes over large areas: differential radar interferometry. J. Geophys. Res. 94, 9183–9191.

Gambolati, G., Freeze, R.A., 1973. Mathematical simulation of the subsidence of Venice. 1. Theory. Water Resour. Res. 9 (3), 721–733.

Gambolati, G., Gatto, P., Freeze, R.A., 1974a. Mathematical simulation of the subsidence of Venice. 2. Results. Water Resour. Res. 10 (3), 563–577.

Gambolati, G., Gatto, P., Freeze, R.A., 1974b. Predictive simulation of the subsidence of Venice. Science 183 (4127), 849–851.

Gambolati, G. (Ed.), 1998. CENAS-Coastline Evolution of the Upper Adriatic Sea due to Sea Level Rise and Natural and Anthropogenic Land Subsidence. Kluwer Academic Pub., Dordrecht, 344 pp.

Gambolati, G., Teatini, P., Ferronato, M., 2005. Anthropogenic land subsidence. In: Anderson, M.G. (Ed.) Encyclopedia of Hydrological Sciences, vol. IV,. Chapter 158 J. Wiley.

Gambolati, G., Teatini, P., Ferronato, M., Strozzi, T., Tosi, L., Putti, M., 2009. On the uniformity of anthropogenic Venice uplift. Terra Nova 21, 467–473.

Gentilomo, M., Cecconi, G., 1997. Flood protection system designed for Venice. Hydropower Dams 2, 46–52.

Ghezzo, M., Guerzoni, S., Cucco, A., Umgiesser, G., 2010. Changes in Venice Lagoon dynamics due to construction of mobile barriers. Coast. Eng. 57, 694–708.

Handwerk, B., 2012. Injections could lift Venice 12 inches, study suggests. Nat. Geog. January 12, 5 pp.

Harleman, D.R.F., 2002. Saving Venice from the sea. J. Hydraul. Res. 40 (6), 81–85.

Kosloff, D., Scott, R.F., Scranton, J., 1980. Finite element simulation of Wilmington oil field subsidence: II. Nonlinear Modelling. Tectonophysics 70, 159–183.

Lorenzetti, G., 1926. Venice and its Lagoon, Historical – Artistic Guide. Lint Publ., Trieste. 1025 pp.

Marchesi, P., 1978. Il forte si Sant'Andrea a Venezia. Stamperie di Venezia Publ., Venice. 113 pp.

Massonnet, D., Rossi, M., Carmona, C., Adragna, F., Peltzer, G., Feigl, K., et al. 1993. The displacement field of the Landers earthquake mapped by radar interferometry. Nature 364, 138–142.

Montanelli, I., Gervaso, R., 1965. L'Italia dei Secoli Bui (Middle Ages to 1000). Rizzoli Publ., Milan. 524 pp.

Montanelli, I., Gervaso, R., 1966. L'Italia dei Comuni (1000–1250). Rizzoli Publ., Milan. 435 pp.

Montanelli, I., Gervaso, R., 1967. L'Italia dei Secoli d'Oro (1250–1492). Rizzoli Publ., Milan. 445 pp.

References

Montanelli, I., Gervaso, R., 1968. L'Italia della Controriforma (1492–1600). Rizzoli Publ., Milan. 589 pp.

Montanelli, I., Gervaso, R., 1969. L'Italia del Seicento (1600–1700). Rizzoli Publ., Milan. 511 pp.

Montanelli, I., Gervaso, R., 1970. L'Italia del Settecento (1700–1789). Rizzoli Publ., Milan. 702 pp.

Nosengo, N., 2003. Venice floods: save our city!. Nature 424 (6949), 608–609.

Pirazzoli, P.A., 2002. Did the Italian government approve an obsolete project to save Venice. EOS Trans. 83 (20), 217–223.

Pratt, W.E., Johnson, D.W., 1926. Local subsidence of the Goose Creek oil field. J. Geol. XXXIV (7), 577–590.

Sethre, J., 2003. The Souls of Venice. McFarland, Jefferson, N.C.

Strozzi, T., Teatini, P., Tosi, L., 2009. TerraSAR-X reveals the impact of the mobile barrier works on Venice coastland stability. Remote Sens. Environ. 113 (12), 2682–2688.

Teatini, P., Gambolati, G., Tosi, L., 1995. A new 3-D non-linear model of the subsidence of Venice. In: Barends, F.B.J., et al. (Eds.), Fifth Int. Symp. on Land Subsidence, The Hague, pp. 353–361. IASH Publication No. 234.

Teatini, P., Ferronato, M., Gambolati, G., Baù, D., Putti, M., 2010. Anthropogenic Venice uplift by seawater pumping into a heterogeneous aquifer. Water Resour. Res. 46, W11547. http://dx.doi.org/10.1029/2010WR009161.

Teatini, P., Gambolati, G., Ferronato, M., Settari, T., Walters, D., 2011a. Land uplift due to fluid injection. J. Geodyn. 51, 1–16.

Teatini, P., Castelletto, N., Ferronato, M., Gambolati, G., Tosi, L., 2011b. A new hydrogeological model to predict anthropogenic Venice uplift. Water Resour. Res. 47, W12507. http://dx.doi.org/10.1029/2011WR010900.

Teatini, P., Tosi, L., Strozzi, T., Carbognin, L., Cecconi, G., Rosselli, R., et al. 2012. Resolving land subsidence within the Venice Lagoon by persistent scatterer SAR interferometry. Phys. Chem. Earth 40–41, 72–79.

Tosi, L., Carbognin, L., Teatini, P., Strozzi, T., Wergmüller, U., 2002. Evidence of present relative land stability of Venice, Italy, from land, sea, and space observations. Geophys. Res. Lett. 29 (12), 1562.

Tosi, L., Teatini, P., Brancolini, G., Zecchin, M., Carbognin, L., Affatato, A., et al. 2012. Three-dimensional analysis of the Plio-Pleistocene seismic sequences in the Venice Lagoon (Italy). J. Geol. Soc. XX. http://dx.doi.org/10.1144/0016-76492011-093.

Umgiesser, G., Canu, D.M., Cucco, A., Solidoro, C., 2004. A finite element model for the Venice Lagoon. Development, set up, calibration and validation. J. Mar. Syst. 51, 123–145.

Vasco, D.W., Rucci, A., Ferretti, A., Novali, F., Bissell, R.C., Ringrose, P.S., et al. 2010. Satellite-based measurements of surface deformation reveal fluid flow associated with the geological storage of carbon dioxide. Geophys. Res. Lett. 37. L03303. http://dx.doi.org/10.1029/2009GL041544.

Zorzi, A., 1979. La Repubblica del Leone – Storia di Venezia. Rusconi Publ., Milan. 737 pp.

Printed and bound by CPI Group (UK) Ltd, Croydon, CR0 4YY
11/06/2025
01899189-0007